Sustainable Civil Infrastructures

Sustainable Civil Infrastructures (SUCI) is a series of peer-reviewed books and proceedings based on the best studies on emerging research from all fields related to sustainable infrastructures and aiming at improving our well-being and day-to-day lives. The infrastructures we are building today will shape our lives tomorrow. The complex and diverse nature of the impacts due to weather extremes on transportation and civil infrastructures can be seen in our roadways, bridges, and buildings. Extreme summer temperatures, droughts, flash floods, and rising numbers of freeze-thaw cycles pose challenges for civil infrastructure and can endanger public safety. We constantly hear how civil infrastructures need constant attention, preservation, and upgrading. Such improvements and developments would obviously benefit from our desired book series that provide sustainable engineering materials and designs. The economic impact is huge and much research has been conducted worldwide. The future holds many opportunities, not only for researchers in a given country, but also for the worldwide field engineers who apply and implement these technologies. We believe that no approach can succeed if it does not unite the efforts of various engineering disciplines from all over the world under one umbrella to offer a beacon of modern solutions to the global infrastructure. Experts from the various engineering disciplines around the globe will participate in this series, including: Geotechnical, Geological, Geoscience, Petroleum, Structural, Transportation, Bridge, Infrastructure, Energy, Architectural, Chemical and Materials, and other related Engineering disciplines.

SUCI series is now indexed in SCOPUS
and EI Compendex.

More information about this series at http://www.springer.com/series/15140

Jose Neves · Bitang Zhu ·
Paulus Rahardjo
Editors

Advanced Geotechnical and Structural Engineering in the Design and Performance of Sustainable Civil Infrastructures

Proceedings of the 6th GeoChina International Conference on Civil & Transportation Infrastructures: From Engineering to Smart & Green Life Cycle Solutions – Nanchang, China, 2021

Springer

Editors
Jose Neves
Department of Civil Engineering,
Architecture & Georesources
Instituto Superior Técnico, Universidade
de Lisboa
Lisbon, Portugal

Paulus Rahardjo
Parahyangan Catholic University
Bandung, Indonesia

Bitang Zhu
School of Civil Engineering
and Architecture
East China Jiao Tong University
Nanchang, China

ISSN 2366-3405 ISSN 2366-3413 (electronic)
Sustainable Civil Infrastructures
ISBN 978-3-030-80154-0 ISBN 978-3-030-80155-7 (eBook)
https://doi.org/10.1007/978-3-030-80155-7

This Springer imprint is published by the registered company Springer Nature Switzerland AG
The registered company address is: Gewerbestrasse 11, 6330 Cham, Switzerland

Introduction

Innovation in geotechnical and structural engineering plays a pivotal role in formulating and promoting smart and green life cycle solutions of civil and transportation infrastructures. This conference is an international forum for discussion and sharing of experiences between researchers and professionals. The present volume includes 8 technical papers of the conference related to advanced geotechnical and structural engineering in the design and performance of sustainable civil infrastructures.

This publication would not have been possible without the efforts of the anonymous reviewers, working in conjunction with the authors, to shape these technical papers to be most useful to practitioners and designers. Each paper received at least two full reviews with the supervision of volume editors. The guidance of Dr. Dar Hao Chen must also be noted for coordinating this volume and contributing to the quality of papers contained in the proceedings of this 6th GeoChina International Conference on Civil & Transportation Infrastructures: From Engineering to Smart & Green Life Cycle Solutions, Nanchang, China, 2021.

Contents

About the Editors

Prof. José Neves is Assistant Professor of the Department of Civil Engineering, Architecture and Georesources, Faculty of Engineering (Instituto Superior Técnico), University of Lisbon, and Senior Member of the Civil Engineering Research and Innovation for Sustainability (CERIS) (ORCID ID: 0000-0002-7131-7967). He completed the Ph.D. in civil engineering at the Technical University of Lisbon and the MSc in soil mechanics at the New University of Lisbon. He is Member of the Technical Committee on Transportation Geotechnics (TC202) of the International Society for Soil Mechanics and Geotechnical Engineering (ISSMGE) and the Academy of Pavement Science and Engineering (APSE). He has authored, co-authored and edited over one hundred of scientific journal papers, books, chapters and conference papers. He is Member of the Editorial Board of Springer Nature Applied Sciences Journal.

Prof. Bitang Zhu is Director of the Engineering Research & Development Centre for Underground Technology of Jiangxi Province, China. He is Chartered Professional Engineer with 10-year experience in engineering consultancy on tunnelling. His research interests are in tunnelling, piling, deep excavation engineering and technology on real-time remote monitoring. He has been the principal investigators or technical investigators for 2 research grants funded by National Natural Science Foundation of China and 18 industry-jointed projects. He was awarded both his bachelor's and master's degrees at the Wuhan University and doctoral degree at the Tongji University, China, and has authored/co-authored more than 50 journal and conference papers and authored one book on laterally loaded piles. He is Fellow of the Engineers Australia.

Prof. Paulus Rahardjo completed undergraduate study at Universitas Katolik Parahyangan (UNPAR) and since then has been Faculty Member at the university. He pursued graduate study in highway engineering at Bandung Institute of Technology (ITB) and then master's degree and Ph.D. degree from Virginia Tech, USA. He has been actively engaged in teaching, research as well as hundreds of geotechnical consultancy. He works for design and advising clients on many geotechnical problems including building foundations, highways, tunnels, bridges, jetty and wharfs, dams, coal mining, etc. His specialties with intense experience in research and practice are in the field of in situ testing and landslides or slope protections. He has written more than 200 articles/papers, research reports and manuals. He has served the university as Department Head, Vice Dean of Faculty of Engineering, Director of the Graduate Programme and Vice Rector for Academic Affairs. Currently, he is Coordinator of Geotechnical Engineering Division and Director of Research Center for Infrastructure and Urban Development. His affiliations include the Indonesian Geotechnical Society (HATTI), American Society of Civil Engineers in the Geo-Institute, the Indonesian Experts on Disasters (IABI) and Board Representative of International Consortium on Landslides (ICL), currently responsible as Head of Deep Foundation Research Institute at Universitas Katolik Parahyangan, Bandung, Indonesia.

Resiliency of Power Grid Infrastructure Under Extreme Hazards - Observations and Lessons Learned from Hurricane Maria in Puerto Rico

Shen-En Chen[1](✉), Miguel A. Pando[2], Agustín A. Irizarry[3], Yamilka Baez-Rivera[4], Wenwu Tang[5], and Yenki Ng[1]

[1] Department of Civil and Environmental Engineering, University of North Carolina at Charlotte, Charlotte, NC 28223, USA
{schen12,yng3}@uncc.edu

[2] Department of Civil, Architectural, and Environmental Engineering, Drexel University, Philadelphia, PA 19104, USA
Miguel.a.pando@drexel.edu

[3] Department of Electrical and Computer Engineering, Universidad de Puerto Rico Mayagüez, Mayagüez, PR 00681, USA
agustin.irizarry@upr.edu

[4] Department of Engineering Technology and Construction Management, University of North Carolina at Charlotte, Charlotte, NC 28223, USA
ybaezriv@uncc.edu

[5] Center for Applied GIScience, Department of Geography and Earth Sciences, University of North Carolina at Charlotte, Charlotte, NC 28223, USA
WenwuTang@uncc.edu

Abstract. On the morning of Wednesday, September 20, 2017, Hurricane Maria made landfall near the southeastern town of Yabucoa, Puerto Rico (PR), as a powerful Category 4 storm on the Saffir-Simpson hurricane wind scale. Hurricane Maria moved diagonally across the island with sustained winds of 249 km/h and is considered the worst storm to hit PR in over 80 years. Hurricane Maria arrived only two weeks after Hurricane Irma; this prior hurricane had passed just north of the island, ensuing heavy rainfall throughout the island, and leaving about one million residents without power. The scale of Hurricane Maria's destruction was even more devastating, causing as much as $95 billion in damages. Electricity was cut off for 100% of the island, and most residents suffered from limited access to clean water and food. Puerto Rico's power outage was, by far, the most severe in United States history in terms of total customer-hours lost.

This paper describes the event timeline and summarizes reconnaissance observations by the authors as part of an NSF RAPID project to document the infrastructure damages of the power grid in PR. Extreme natural disasters associated with climate change have increased in frequency in recent years, resulting in significant impacts on local economies and drastic increases in global disaster expenditures. Many of these climate events, including wind storms, ice storms, hurricanes, floods, landslides, and tornadoes, are directly affecting the people, infrastructure, economies, and the natural environment. A storm effect correlation analysis was performed on a 10 km × 10 km grid on the island and the most damaging storm effect is identified. The field information and data presented in

J. Neves et al. (Eds.): GeoChina 2021, SUCI, pp. 1–17, 2021.
https://doi.org/10.1007/978-3-030-80155-7_1

this paper provide insight to help the engineering community adapt and improve design and construction practices to improve resiliency of our infrastructure and lifelines.

Keywords: Hurricane Maria · Puerto Rico · Power grid · Physical resilience

1 Introduction

The 2017 hurricane season was proven to be a unique season for the US with three distinct hurricanes: Harvey, Irma, and Maria totaling $265 billion, with each costing about $125 billion, $50 billion, and $90 billion, respectively (NOAA 2019a). Although it was not the most economically damaging, hurricane Maria was one of the worst natural disasters on record for the US. The damaging effects of Maria manifested on the island of Puerto Rico in the form of power outage: Electricity was cut off for 100% of the island, and most residents suffered from limited access to clean water and food. Puerto Rico's power outage was, by far, the most severe in United States history in terms of total customer-hours lost (NOAA 2019b).

On the morning of Wednesday, September 20, 2017, Hurricane Maria made landfall near the southeastern town of Yabucoa, Puerto Rico, as a powerful Category 4 storm on the Saffir-Simpson hurricane wind scale. Hurricane Maria moved diagonally across the island with sustained winds of 249 km/h and is considered the worst storm to hit Puerto Rico in over 80 years. On record, Hurricane Maria peaked as a category 5 hurricane, topping at maximum wind speed of 278 km/h (Pasch et al. 2019). Associated with the strong wind forces were torrential rainfalls, foliage losses (deforestation), flooding and storm surges. As a result, the storm effects from Maria destroyed 80% of Puerto Rico's utility poles and almost all transmission lines (USDOE 2018). Figure 1 shows the damaged power structures at different parts of the island. It was reported that only half of the island's power was restored by the end of 2017, and 65 percent was restored only by the end of January 2018 (Pasch et al. 2019). It took approximately a whole year for complete power restoration (Campbell 2018). Also shown in Fig. 1 are the seven regions of the main island power grid: Mayagüez, Arecibo, Ponce, Bayamón, San Juan, Carolina and Caguas.

Grid resilience is the measure of a power system's ability to recover from the effects of a severe storm, which can be enhanced through grid hardening strategies including selection of stronger power structures, mix energy production, energy storage, smart grid and better power infrastructure management techniques, etc. (Reed et al. 2009; NAP 2017). Forensic investigation generates important information that can contribute to the risk and gap analyses for infrastructure resilience and further enhance system performance during future storms (Habibian and Minaei 2018). Understanding of the actual failure modes would help in revealing how the power grid has performed in the extreme storm and how it can be strengthened against future disasters.

For past significant storms, forensic investigations of damaged power systems have been conducted (Reed et al. 2010; Winkler et al. 2010; Dai et al. 2017). As the frequency of strong hurricane events has been projected to increase in the very near future in the form of climate change (Krishna 2009; Nigam and Guan 2011; McDonald 2011; Vickery and

Lavelle 2012; Emanuel 2013), such studies will become more critical. This is especially true for island states such as Puerto Rico where the power grid is characterized as isolated and with vertically integrated utility monopolies.

This paper describes the event timeline and summarizes reconnaissance observations by the authors as part of an NSF RAPID grant to document the infrastructure damage of the power grid in PR. The storm effects are presented in 10 km × 10 km grid plots and the dominant storm effects are identified using a correlation factor analysis. The key lessons learned that may be valuable towards an improved grid system resiliency under these types of extreme hazards are summarized.

2 Maria's Path and Impacts

Hurricane Maria originated off the west coast of Africa as a tropical disturbance on September 12, 2017 and made its way west towards the Caribbean islands and strengthened into a hurricane with wind speed at about 185 km/h (NOAA, 2019b). On September 19th, 2017 Maria made its initial landfall on the island of Dominica with a sustained maximum wind speed of about 269 km/h and a minimum central pressure of 92 kPa. After Dominica, Maria continued its travel path and made way towards the island of Puerto Rico. Before reaching Puerto Rico, the hurricane peaked at a wind intensity of 278 km/h with a central pressure of 91 kPa. At the point of contact, in the southeast coast of the island, the maximum wind intensity weakened to about 249 km/h, just below the threshold of a category 5 hurricane. Maria made its way through Puerto Rico from the southeast corner to the northwest corner with a duration of about several hours. After passing through Puerto Rico, the hurricane weakened again to about 176 km/h wind intensity. Maria kept its consistent travel path moving northwest until it changed its direction northward on September 22. While still maintaining a hurricane intensity, By September 27, Maria changed its direction again and finally dissipated on October 2, 2017. Figure 2 shows the travel path of Maria as it was making its way through the island. Also shown in figure is the power grid and the relative power loss during Maria. As shown, the significant relative power losses are on the east side of the island (most populated region).

Silva-Tulla et al. (2018) described in detail various storm effects during and after the hurricane in Puerto Rico including strong winds, storm surges, rainfall and flooding, coastal and river erosions, and landslides, resulting in massive damages to structures and infrastructures. There were also significant trees down due to strong wind and torrential rain (Bessette-Kirton et al. 2019). These can all be the causes of some or majority of the power structures to fail.

To demonstrate impacts of Maria, Figs. 3, 4, 5, 6, 7, 8 and 9 show the individual storm effects (rainfall, landslide, flooding, storm surge, coastal erosion, maximum wind speeds and deforestation, respectively) on the island - The effects are condensed into 10 km × 10 km grids and presented along with the major power grid on the island. As shown in the figures, the severe rainfall (Fig. 3) during the hurricane caused major occurrences of landslides throughout the island. According to Bessette-Kirton et al. (2019), hurricane Maria triggered more than 40,000 landslides in at least three-fourths of Puerto Rico's 78 municipalities. This was caused by the elevated pore-water pressure within the soil as a result of the heavy rainfall. Figure 4 shows the extent of landslides and shown most severe slides are populated in the central mountain range.

Heavy rainfall also resulted in localized flooding, which is shown in Fig. 5. Some of the flooding would eventually resulted in infrastructure (i.e. bridges and dams) damages and induced local ground subsidence. The strong wind force reaching the island resulted in storm surges around the coastal regions (Fig. 6). It was reported that Puerto Rico experienced a maximum inundation levels of 1.83 m to 2.74 m above sea level resulting from a combined effects of storm surge and tide (Pasch et al. 2019). The storm surge would trigger coastal erosions, which occurred at various locations around the island (Fig. 7).

Finally, one of the significant effects of the strong wind force is the result of deforestation in the forms of trees down and loss in foliage on the tropical island. Figure 8 shows the strong wind force effect (represented by maximum wind speeds experienced throughout the island), which formed a band diagonal through the island. The most significant effect of foliage change is shown, however, to focus mostly on the south-western part of the central mountain range in Puerto Rico (Fig. 9).

3 Investigation of Damaged Structures in Puerto Rico

Following hurricane Maria, on-the-ground data collection for structural assessment was performed to determine the damage severity of residential structures using a rating system (Chen et al. 2016). The rating system consists of defining a damaged structure as either minor, moderate, or major. The rating technique was previously devised during disaster assessment of another significant storm event, super typhoon Haiyan (Chen et al. 2015). The data were captured by a field team that was made up of both civil and electrical engineers and was sent to the Puerto Rico in early May 2018, eight months after the storm (Chen et al. 2020). The team traveled throughout the island and captured photographs of damaged residential structures. The photographs also include a wide range of the island's disaster damages such as topography, roadways, and different types of infrastructures (transmission, bridges, commercial and historical buildings, etc.). Over 10,000 photographs and 7,000 satellite images were collected. Figure 10 shows images of damaged buildings that were captured during the field study indicating different types of structural damages.

The purpose of the structural assessment was to provide a comprehensive appreciation of the damage extents on the island. The damage structures were identified by their geographical locations and assembled using GIS software for geospatial analysis. The damaged states are presented by quantifying the distribution by each power region: As shown in Fig. 11, the damage rating distributions for each power region are displayed. The results indicate a significant number of structures are rated as minor or moderately damaged. There are more moderately damaged structures in San Juan metropolitan areas (75.7%), Bayamon (67.4%), Mayaguez (64.9%), Ponce (64.4%) and Caguas (63.2%). Most of the moderately damaged structures are due to wind-induced roof damages. The damage evaluation results are also summarized in Table 1. These damages were influenced by many possible storm effects. To delineate the damage effects, the most significant impacts from each storm effects are superimposed onto the power grid (Fig. 2) and the results (Fig. 12) shows the graphical representation of the storm damaged power grid, which will be discussed in the following section.

4 Puerto Rico's Power Grid Situation After Maria

The power supply scenario prior to Maria is first described: The power grid on the island of Puerto Rico is managed by the Puerto Rico Electric Power Authority (PREPA). PREPA's power grid on the main Puerto Rico island has more than 3,862 km of transmission lines (230 and 115 kV), fifty-one 115 kV transmission centers, 283 substations (38 kV), and over 48,280 km of distribution lines (13.2, 8.32, 7.2, and 4.16 kV). The main power generation is installed on the southern part of the island with a total generation capacity of 3,443 MW. Since 70% of the electricity demand occurs in the northern coastal areas (specifically in the northeast), there is significant reliance on the power grid to deliver energy from the southern power plants to the north. Because of the presence of the central mountain range, the power grid is looped around the island and with some very high voltage transmission lines riding along the mountain ridges.

The supporting structures for the power grid in Puerto Rico include concrete, wood and steel poles and large truss towers. The truss structures include lattice structures, guyed trusses, H-frames, as well as single column delta structures. Wood poles are the predominant structure for power distribution lines.

Damaged power systems throughout Puerto Rico were proven to be a substantial problem due to Maria - The damage modes of power delivery structures indicated predominantly failure by wind overload resulting in concrete pole buckling, conductors down, localized torsional buckling of steel members, and loss of equipment such as insulators, ground lines, jumpers and transformers. Furthermore, foundation failures happened to both pole-like structures as well as truss structures as a result of landslides or debris flow. Figure 13 shows failures of both pole and truss structures. In several cases, domino-like failure modes are demonstrated by multiple structural failures along a power line. For example, a 115 kV line in Humacao has seen 14 structures experienced significant damages and seven lattice structures completely knocked down. Detailed review of the failure modes indicated that several of the single point supported lattice structures including guyed rigid Y transmission towers, H-frame structures and single column structures have collapsed under high winds, indicating that such structures are not able to resist the pushover from the strong wind forces of Maria.

From the electric system faulting perspective, power structure failures during the strong storm event can lead to different electric faulting in the grid system from simple flickering, voltage swell and sag, to complete power outage. The power outage was not fully recovered until more than 227 days with San Juan region first to fully recover and Caguas the last to fully recover (Kwasinski et al. 2019). According to Lluveras (2018), it took eleven months for the entire island to fully restore its electricity.

5 Discussion

5.1 Storm Effect Correlations

To understand the impacts of Maria, current study focuses on storm effects from Maria including rainfall, landslides, flooding, storm surges, coastal erosion, wind force and deforestation. As shown in Fig. 12, several areas are exposed to multiple storm effects – a clear indication of multiple-hazards within a single hurricane. For example, power regions including Ponce, Arecibo and Caguas have experienced significant wind forcing, landslides, deforetations and flooding. Arecibo and Caguas also suffered from extensive storm surges – however, the transmission lines are far enough from the coastlines, it is assumed that the effects due to storm surges are deemed insignificant for power transmission structures. Bayamon and San Juan, on the other hand, have experienced predominantly deforestation and rainfall.

To further establish the criticality of each storm effect, a correlation factor analysis is performed to determine the most critical effects due to Maria: The correlation factor analysis uses normalized values (0 to 1) from each storm effect (Figs. 3, 4, 5, 6, 7, 8 and 9) and use the following equation to establish the causal-effect relation:

$$Correlation = abs(Cause_i - Effect_j) \tag{1}$$

where i = individual cause and j = individual effect. The storm effects are distinguished into causes (wind forces and rainfall) and effects (deforestation, landslide, coastal erosion, flooding and storm surge). The rationale being: The strong wind forces would result in effects including deforestation, coastal erosion and storm surge and the torrential rainfall would result in effects including landslide and flooding. Equation 1 is computed for each grid element and is computed for the entire island. By taking the absolute of each value, we ignored the positive and negative aspects of the correlation. Further, for the entire island, a total value is computed representing the overall effect. For a strong correlation, the total value would result in a lower number. For example, if we compute the maximum value for the self correlation (auto-correlated) of the wind force effect to itself, the total value will be 0.

Figure 14 shows the different correlation analyses performed: a) maximum wind versus deforestation, b) maximum rainfall versus flooding, and c) maximum rainfall versus landslides. Storm surge and coastal erosion effects are not considered in the study because they are limited to only coastal regions. Again, the smaller numbers represent stronger correlations. For example, Fig. 14a) shows the correlation between wind force and deforestation, and indicates that the maximum wind effects on deforestation actually is in the south-western part of the island (despite the fact that the maximum wind force traveled diagonally across the island). Figure 14b) shows significant flooding focused on the southwest corner of the island away from the central mountain range, which has several localized flooded zones. Landslides have been distributed throughout the island and Fig. 14c) shows that the critical areas are on the south side of the mountain range.

To determine if there is a correlation between strong winds and rainfall using the spatial distribution, Fig. 14d) shows the correlation factor analysis between the two causal effects. Figure 14d) indicates that the combined effects are most significant near the northeastern corner of the island and shows the diagonal distribution consistent with the storm path.

Table 2 shows the total values of the correlation factor analyses and the most significant effects are wind force vs. deforestation (39) and rainfall vs. landslide (38.9). The effects of rainfall on coastal erosion and flooding have less storm effects with total values of 54.04 and 54.38, respectively.

6 Power Structure Hardening Recommendations

The above storm effect analysis indicates that the most significant effects from Maria are strong winds and landslides. From the structural damages studied, we observed that significant number of structures were damaged due to strong winds. Landslides resulted in less structures damaged but the damages are more severe. Thus, structural damage observations confirmed with the results from the correlation analysis. With observations determined through the correlation of storm events in relation to damaged structures, possible solutions can be generated to help with the decision-making process for grid hardening strategies.

It should be noted that resiliency is more than just lessening the likelihood of such occurrences; it is also about limiting the scope and impact of such occurrences, restoring power rapidly afterwards, and learning from these experiences to better deal with events in the future (NAP 2017).

Based on observations made from the failed transmission structures (data shared by PREPA), it is recommended that grid hardening should include replacing the existing transmission and pole structures with structure types that can withstand higher intensity wind loads. Examples would be to replace the guyed, single support structures with four-legged transmission structures (Archana et al. 2015). Single point guy-supported structures lack lateral stability to withstand extreme wind conditions. Using four-legged tower structures, a more stable foundation can help reduce the amount of potential sways of the structure. The wider base is also a tradeoff to the demand of larger right-of-way of guy anchored foundations. Panchal et al. (2016) suggested that four-legged transmission towers are 20% more economical and with 38% less deflection than three-legged towers.

Also, round/tubular pole structures should be used instead of square poles (ASCE 2012). These types of structures are advantageous to withstanding intense wind and rainfall because of its smaller wind profile and flexible rigidity as compared to that of rectangular structures. Tubular structures also are less likely to have torsional effects and can allow for more tip deflection than rectangular structures.

Because Puerto Rico has many forested areas, vegetation management around transmission and distribution structures is another approach that can significantly improve the damaging effects to the power structures.

The biggest concern moving forward is the likely occurrence of another massive storm that can possibly cause the same impact as hurricane. Climate change can attribute to such increase in hurricane intensity because of warmer sea surface temperatures and the rise of sea level (Rezaei et al. 2016). With potential increase occurrences of storm events as intense as hurricane Maria, it is also recommended to find solutions for reducing carbon emission or greenhouse gas increase.

7 Conclusions

This paper presents the results of a reconnaissance mission to Puerto Rico that was funded by NSF (Rapid Project) to assess damage of electrical power grid in Puerto Rico due to the 2017 hurricane Maria. The study focuses on structural damages and by studying the storm effects (including wind force, rainfall, flooding, deforestation, landslide, the contributions from different storm effects were delineated. The storm effects are presented in 10 km × 10 km grids for the five storm effects considered. Such delineation can help determine strategies to harden the power structures and enhance resilience against future storms.

The following conclusions are made:

- From the collected images of damaged residential structures, majority of them were considered minorly or moderately damaged.
- Correlation factor analysis results indicated that the strong wind induced deforestation and torrential rainfall induced landslides are both strongly correlated and contributed to most of the structural damages observed in the field study.
- There were less contributions to structural damages from storm surges, coastal erosion and flooding.
- Investigations of damaged power structures of downed pole structures and truss structures indicated poor structural types that are inherently weak against the unprecedented storm event.
- It is recommended to use four-legged transmission towers and tubular pole structures where necessary and permissible for the hardening of the power grid in Puerto Rico.
- It is also recommended that pole structures should be tubular in shape and preferably stronger material types.

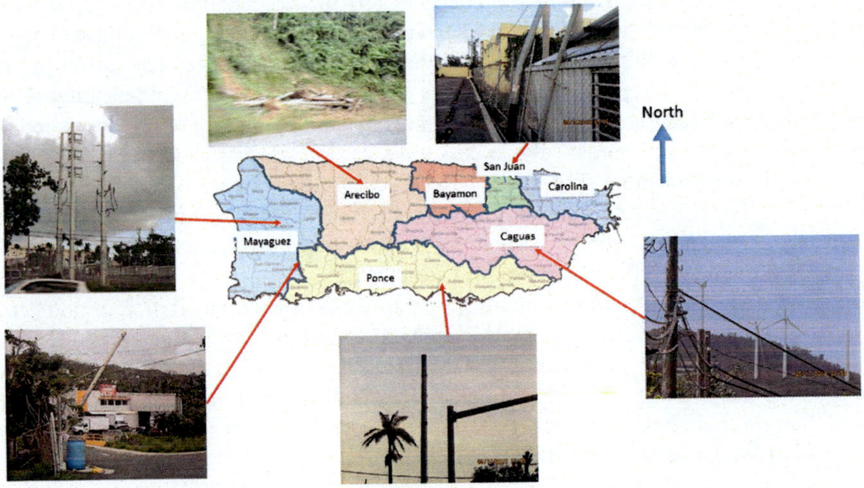

Fig. 1. Example of damaged structures and PR power regions. (Photo Credit: PREPA)

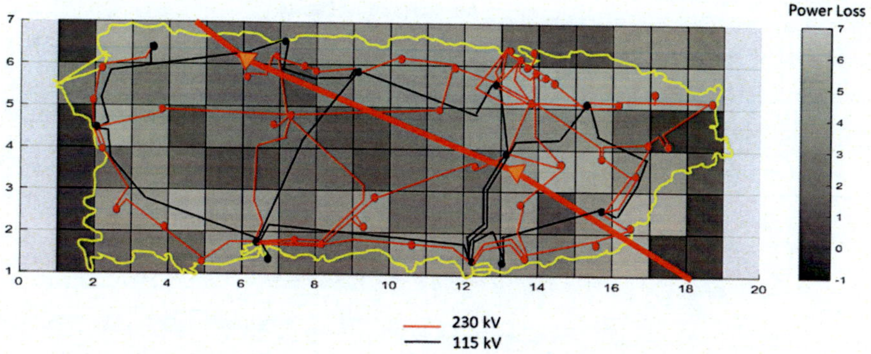

Fig. 2. Travel path of hurricane Maria across Puerto Rico and island power loss.

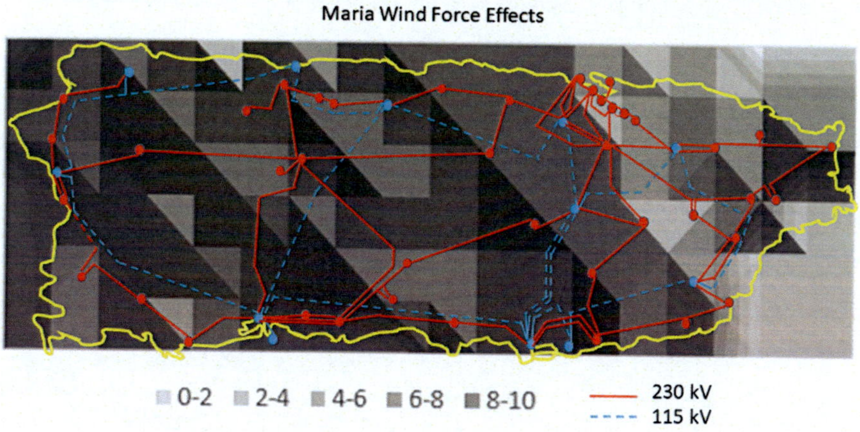

Fig. 3. Severity of storm effect (10 km × 10 km): Rainfall.

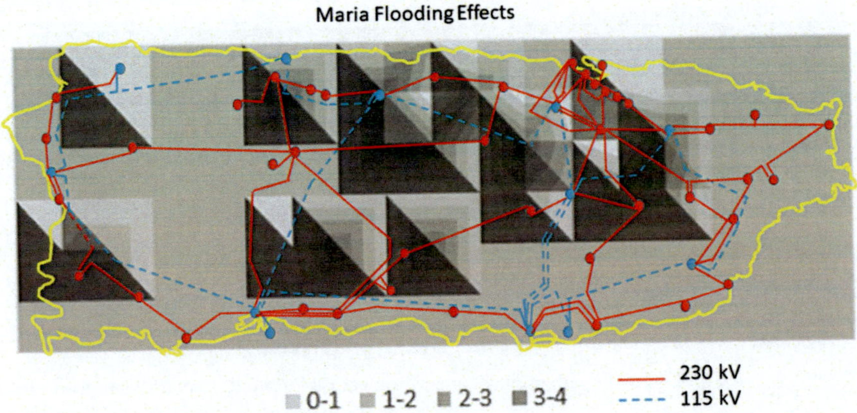

Fig. 4. Severity of storm effect (10 km × 10 km): Landslides.

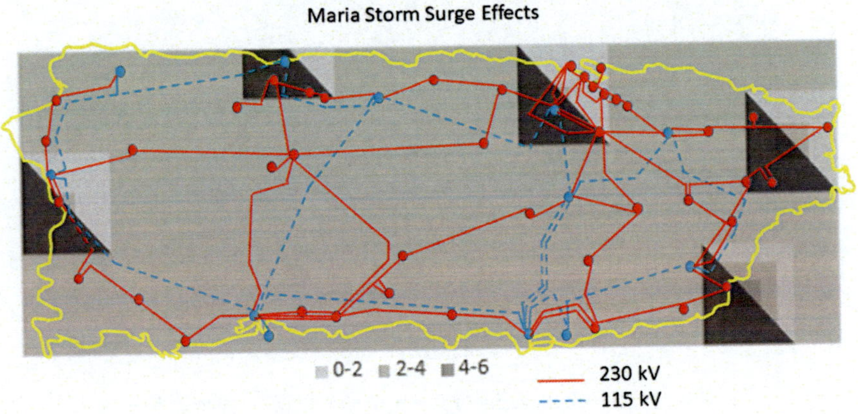

Fig. 5. Severity of storm effect (10 km × 10 km): Flooding.

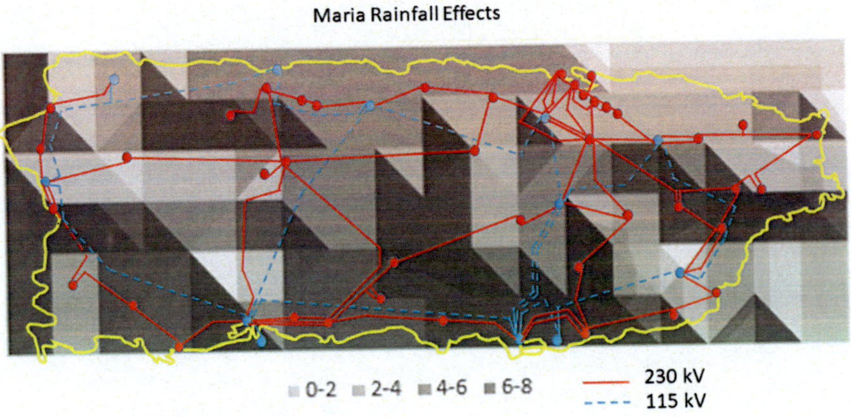

Fig. 6. Severity of storm effect (10 km × 10 km): Storm surge.

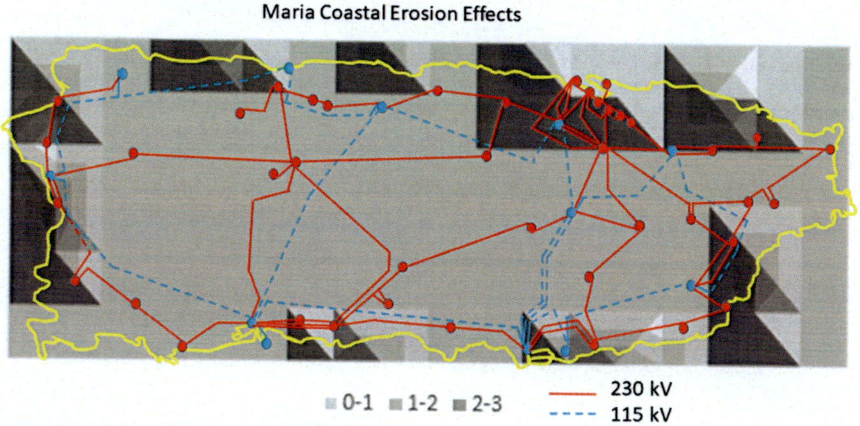

Fig. 7. Severity of storm effect (10 km × 10 km): Coastal erosion.

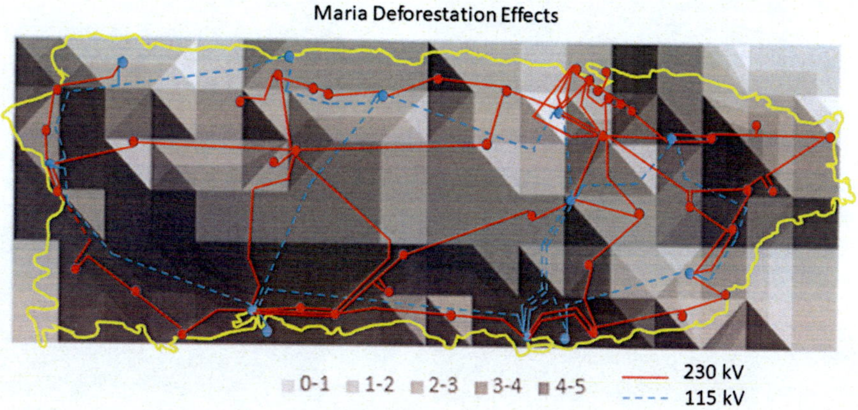

Fig. 8. Severity of storm effect (10 km × 10 km): Wind force.

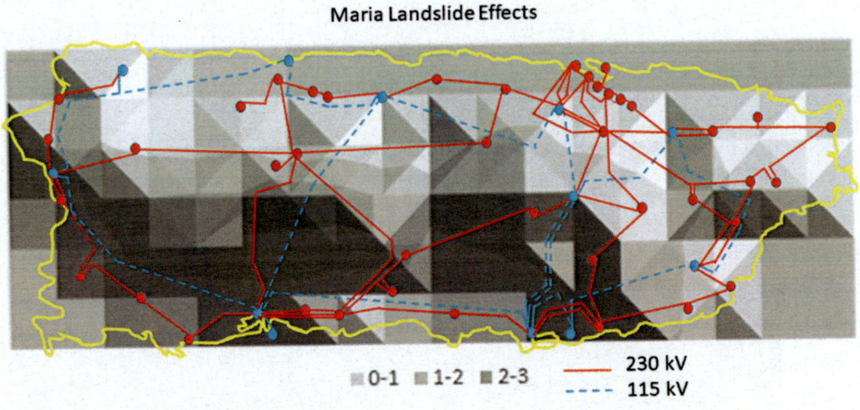

Fig. 9. Severity of storm effect (10 km × 10 km): Deforestation.

Fig. 10. Photos showing examples of structural damage in PR following hurricane Maria (Photos from NSF Rapid mission).

Fig. 11. Damage rating along each region of Puerto Rico.

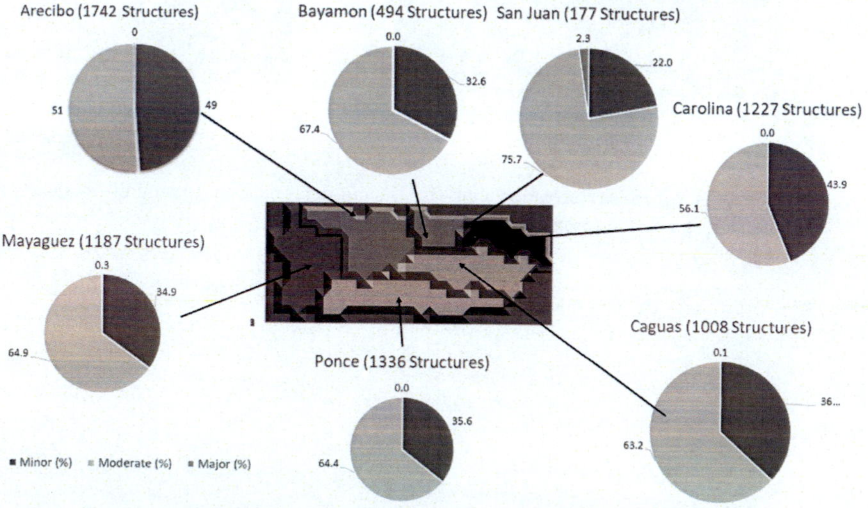

Fig. 12. Storm effect overlapping indicating multi-hazards on the island.

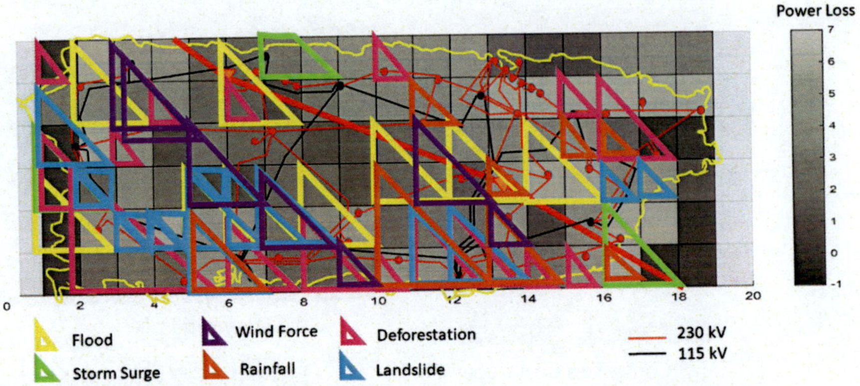

Fig. 13. Various modes of damages of power structures. (Photo Credit: PREPA)

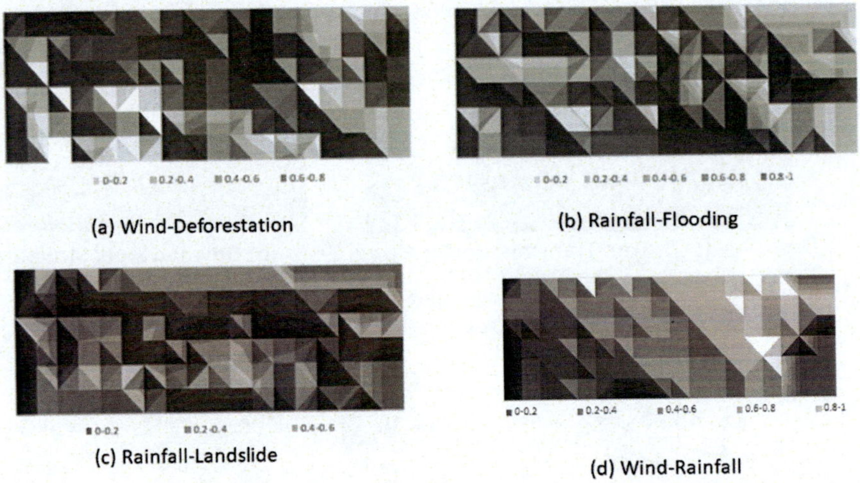

Fig. 14. Storm effect correlation analysis: a) wind-deforestation; b) rainfall-flooding; c) rainfall-landslide; and d) wind-rainfall.

Table 1. Quantitative values of damage severity.

Puerto Rico structural assessment			
Region	Damage severity		
	Minor	Moderate	Major
Arecibo	849	891	2
Bayamón	161	333	0
San Juan	39	134	4
Carolina	539	688	0
Caguas	370	637	1
Ponce	475	861	0
Mayagüez	414	770	3
Total	2847	4314	7

Table 2. Total values from causal-effect correlation analysis

Relation	Total value	Correlation
Wind-deforestation	39	Strong
Rainfall-flooding	54.04	Weak
Rainfall-coastal erosion	54.38	Weak
Rainfall-landslide	38.97	Strong

Acknowledgements. The authors would like to acknowledge the funding received under NSF Grant CMMI-1807813 (NSF Program Director Dr. Anthony Kuh). The research team would also like to extend their gratitude and appreciation to PREPA for supporting this project through access to critical data of power grid responses during Hurricane Maria. In particular, the team would like to acknowledge the help from engineers Efran Paredes, Mireya Rodriguez, Camille Ocasio and Luderis Berrios.

Any opinions, findings, and conclusions expressed in this paper are those of the authors and do not necessarily, reflect the views of the NSF or PREPA.

Data Availability Statement. Some or all data, models, or code generated or used during the study are available in a repository online in accordance with funder data retention policies (http://cybergis.uncc.edu/hurricane).

References

Archana, R., Aswathy, S.K.: Analysis and design of four-legged transmission tower. Int. J. Sci. Res. **2319–7064**, 1026–1030 (2015)

ASCE: Prestressed Concrete Transmission Pole Structures – Recommended Practices for Design and Installation, Manuals and Reports on Engineering Practice No. 123. ASCE, Reston, VA (2012)

Bessette-Kirton, E.K., et al.: Landslides triggered by hurricane Maria: assessment of an extreme event in Puerto Rico. GSA Today Arch. **29**(6), 4–10 (2019)

Buldyrev, S.V., Parshani, R., Gerald Paul, H., Stanley, E., Havlin, S.: Catastrophic cascade of failures in interdependent networks. Nature **464**(7291), 1025–1028 (2010). https://doi.org/10.1038/nature08932

Campbell, A.F.: It Took 11 Months to Restore Power to Puerto Rico after Hurricane Maria. A Similar Crisis Could Happen Again. Vox, 15 August 2018

Chen, S.E., et al.: ASCE Hurricane Haiyan disaster investigation in the Philippines. ASCE J. Perform. Constr. Facil. **29**(4), 02514003 (2015)

Chen, S.E., et al.: Basic structure system rating of post-super typhoon Haiyan structures in Tacloban and East Guiuan, Philippines. J. Perform. Constr. Facil. **30**(5), 04016033 (2016). https://doi.org/10.1061/(ASCE)CF.1943-5509.0000872

Chen, S.E., et al.: Post-hurricane investigations a critical component towards improved grid resiliency-hurricane Maria in the Puerto Rico ASCE J. Perform. Constr. Facil. **34** (2020).https://doi.org/10.1061/(ASCE)CF.1943-5509.0001447

Cuello-Polo, G.A., Irrizarry-Rivera, A.: Power Flow Analysis of Hurricane Maria Impact on Puerto Rico's Electric Power Network, UPRM Final Report, NSF-RAPID: 1807813, Mayagűez, PR (2020)

Dai, K.S., Chen, S.E., Loflin, G., Luo, M.: A framework for holistic designs of power line systems based on lessons learned from super typhoon Haiyan. Sustain. Citites and Soc. **35**, 50–364 (2017). https://doi.org/10.1016/j.scs.2017.08.006

Emanuel, K.A.: Downscaling CMIP5 climate models shows increased tropical cyclone activity over the 21st century. Proc. Nat. Acad. Sci. (PNAS) **110**(30), 12219–12224 (2013)

Habibian, A., Minaei, E.: Achieving lifeline infrastructure resilience through an adaptive and risk-based approach. J. Am. Water Works Assoc. (AWWA) **110**(8), 42–49 (2018)

Krishna, K.M.: Intensifying tropical cyclones over the North Indian Ocean during summer monsoon-global warming. Glob. Planet. Changes **65**, 12–16 (2009)

Kwasinski, A., Andrade, F., Castro-Sitiriche, M.J., O'Neill-Carillo, E.: Hurricane Maria effects on Puerto Rico electrical power infrastructure. IEEE Power Energy Technol. Syst. J. **6**(1), 85–94 (2019)

Lluveras, L.: Puerto Rico Has Not Recovered from Hurricane Maria. Public Radio International, 19 September 2018

McDonald, R.E.: Understanding the impact of climate change on northern hemisphere extra-tropical cyclones. Clim. Dyn. **37**, 1399–1425 (2011)

Nigam, S., Guan, B.: Atlantic tropical cyclones in the twentieth century: natural variability and secular change in cyclone count. Clim. Dyn. **36**, 2279–2293 (2011)

NAP 2017: Enhancing the Resilience of the Nation's Electricity System. The National Academies Press (2017). ISBN 978-0-309-46307-2

NASA: Pinpointing where Lights Went Out in Puerto Rico. The NASA (2017). https://earthobservatory.nasa.gov/images/91044/pinpointing-where-lights-went-out-in-puerto-rico. Accessed 6 Nov 2020

NOAA: Hurricane Costs. Office for Coastal Management, National Oceanic and Atmospheric Administration, 10 July 2019 (2019a). https://coast.noaa.gov/states/fast-facts/hurricane-costs.html. Accessed 6 Nov. 2020

NOAA: Weather Disasters and Costs. Office for Coastal Management, National Oceanic and Atmospheric Administration (2019b). https://coast.noaa.gov/states/fastfacts/weatherdisasters.html. Accessed 20 Jun 2020

Pasch, R.J., Penny, A.B., Berg, R.: National Hurricane Center Tropical Cyclone Report Hurricane Maria (AL152017). National Oceanic and Atmospheric Administration, 14 February 2019 (2019)

Panchal, H.S., Vyas, V.H., Desai, H.G.: Comparative analysis of transmission line tower with different using conventional angle section and closed hollow section. J. Civ. Eng. Environ. Technol. **3**(4), 267–268 (2016)

Peraza, D.B., Coulbourne, W.L., Griffith, M.: Engineering Investigation of Hurricane Damage: Wind versus Water. ASCE, Reston, VA (2014)

Reed, D.A., Kapur, K.C., Christie, R.D.: Methodology for assessing the resilience of networked infrastructure. IEEE Syst. J. **3**(2), 174–180 (2009)

Reed, D.A., Powell, M.D., Westerman, J.M.: Energy supply system performance for hurricane Katrina. J. Energy Eng. **136**(4), 95–102 (2010)

Rezaei, S.N., Chouinard, L., Langlois, S., Legeron, F.: Analysis of the effect of climate change on the reliability of overhead transmission lines. Sustain. Urban Areas **27**, 137–144 (2016)

Silva-Tulla, F., et al.: Geotechnical Impacts of Hurricane Maria in Puerto Rico. GEER Association Report No. GEER-057 (2018). https://doi.org/10.18118/G68083. http://www.geerassociation.org

US DOE: Energy Resilience Solutions for the Puerto Rico Grid, Final Report, US Department of Energy, Washington, D.C. (2018)

Vickery, P.J., Lavelle, F.M.: The effects of warm Atlantic Ocean sea surface temperature on the ASCE 7-10 design wind speeds. In: Jones, C.P., Griffis, L.G. (eds.) Advances in Hurricane Engineering, pp. 13–22. ASCE (2012)

Winkler, J., Dueñas-Osorio, L., Stein, R., Subramanian, D.: Performance assessment of topological diverse power systems subjected to hurricane events. Reliab. Eng. Syst. Saf. **95**, 323–336 (2010)

Methodology that Combines Multi-criteria Methods for Decision-Making, Hierarchical Analytical Process and the Goal Programming, and Their Impact in the Sustainability Evaluation of Hydroelectric Projects in Mexico

José Andrés Gómez Romero[1], Susana Garduño Román[2(✉)], Humberto Marego Mogollón[3(✉)], and María del Rocío Soto-Flores[4(✉)]

[1] Federal Electricity Commission (CFE), Hydroelectric Projects (CPH), Mexico City, Mexico
andres.gomez02@cfe.gob.mx

[2] Postgraduate Studies and Research of the Higher School of Commerce and Administration (ESCA), Santo Tomás Unit of the National Polytechnic Institute (IPN), Mexico City, Mexico
sgarduno@ipn.mx

[3] The International Boundary and Water Commission By the United States and Mexico, Mexico City, Mexico

[4] Higher School of Commerce and Administration (ESCA), Santo Tomás Unit of the National Polytechnic Institute (IPN), Mexico City, Mexico

Summary. Context: Hydroelectricity is a mature and long-lasting technology that has presented environmental and social problems, to face them, hydroelectric sustainability initiatives have been created for more than two decades. In Mexico, hydroelectric projects present environmental and social problems that cause delays during their construction and, sometimes, when they are put into operation. The objective of this research is to report how the combination of multi-criteria decision-making methods, such as the Hierarchical Analytical Process (AHP) and the Goal Programming (GP) could be applied to weight the hydroelectric sustainability criteria, using multi-criteria decision-making methods for a group of experts.

Method: The combination of the multi-criteria methods, the Hierarchical Analytical Process (AHP) and the Programming by Goals (GP) were used to identify the opinion of experts in the planning of hydroelectric projects. From here, the IHA protocol was used to determine the sustainability profiles with a triangulation by methods, using the AHP sensitivity analysis.

Results: The combination of AHP-GP methods was applied to establish the hierarchies of hydroelectric sustainability criteria and sub criteria, as well as a documentary analysis which helped to determine the hydroelectric sustainability profiles of hydroelectric projects. One of the most sustainable hydroelectric projects was "Las Cruces" due to the results obtained through a sensitivity analysis with five scenarios, from which four out five were validated and located it in the first place.

Conclusions: The results obtained indicate that when questioning a heterogeneous group of experts, there were significant agreements between them, despite

J. Neves et al. (Eds.): GeoChina 2021, SUCI, pp. 18–42, 2021.
https://doi.org/10.1007/978-3-030-80155-7_2

they were from different areas of knowledge. Regarding theoretical implications, it was observed that the combination of AHP-GP multicriteria methods and a documentary analysis should be considered through a mixed research method.

Keywords: Heterogeneous groups · Hierarchical analytical process · Goal programming · Hydroelectric sustainability · Decision making

1 Introduction

The combination of the multi-criteria methods, the Hierarchical Analytical Process (AHP) and the Programming by Goals (GP) methods for solving problems related to multiple decisions have been used by several researchers such as Gass and Rapcsák (1998), Schniederjans et al. (1995), Kim et al. (1999), Badri (2001), Yurdakul (2004), Perçin (2006), Aznar and Estruch (2007), Mendoza et al. (2008), Erdem and Göçen (2012), Kozioł (2014), Kambiz et al. (2016), Hamurcu et al. (2017); Özan et al. (2017); and Wichapa and Khokhajaikiat (2017), among others. But none of them have applied this combination, AHP - GP, for a direct application in hydroelectric sustainability as was carried on by the research under analysis. Therefore, this paper pays direct attention to the way AHP - GP combination was implemented on hydroelectricity projects in Mexico. Derived from this, is important to look at the increasing demand of energy in the world and the actions adopted to fulfil the need through hydroelectric sustainability and to improve decision-making around hydroelectric projects.

Hydroelectricity is a source of energy coming from water which has been the most sought resource in the establishment of communities; rivers have supplied electricity to cities and industries through the dams which were built as instruments to divide or distribute water of the rivers since 3,000 B.C., as from the Jordan River (McCully 2004). According to the World Commission on Dams (WRC 2000), the 20th century presented the greatest increase in dam construction. In 1949, derived from economic growth after the Second World War, it led to the construction of 5,000 dams in the world, three quarters of them in industrialized countries. They were the symbol of modernity and of humanity's ability to use and control nature's water resources. Dam construction grew, and reached its peak in the 1970s when, on average on the planet, two to three dams were opened per day (CMR 2000). Dam construction declined since the early 1990s, when the World Bank restricted financing for these projects due to the impact in the environment.

Hydroelectricity is more than a century old, the first one was built in 1880 in England. With this, the electric generator and the hydraulic turbine were developed, with which, at the beginning of the millennium, their demand increased (Salas, s.f.). But the awareness of its impact was after the 1960s. This late awareness was due to the fact that environmental concern began after that time, together with the development of hydroelectric power in the world (Schoijet 1984). Initially, its planning and evaluation were limited to technical parameters and economic analysis; social and environmental impacts were excluded from the evaluation and their role in project selection was secondary (CMR 2000). In 1991, an internal World Bank survey indicated that 58% of projects were planned without considering the downstream environmental effects they would cause (Pottinger, in Atwi and Arrojo 2000).

For Suárez and Peirano (2010), during the construction of these projects, different types of problems arise. Among them were, the movement of terrain, the extraordinary movement of construction materials, merchandise and people, as well as the generation of noise and dust, besides erosion, the collapse of the road system, the construction of transmission lines and access to roads. All these have caused modifications to vegetation in the wild lands, the fauna, the soils, the fishing, the climate, and the towns of the project area. But the greatest impact was on the water reservoir as it permanently floods large areas of land. According to the CMR (2000), it produced between 40 to 80 million displaced by the construction and operation of the dams.

But this operation of dams is a response to the increasing demand for energy on the planet, and especially in Latin America, which nowadays offers the probability of using renewable energy resources. For the International Energy Agency (IEA 2018), in 2012, electricity consumption in this region reached 948,000 GWh, and hydroelectric energy contributed 66% of the total production. In Central America it rose above 90%, since it is a mature and long-lasting technology, with some plants in operation that are over 100 years old (Kumar et al. 2011).

On the other hand, the areas of concern for environmental and social issues are related to the lack of mitigation measures, adequate compensation, monitoring and accumulation of impacts. Similarly, in a successful development of a hydroelectric project, the importance of attention to human rights must be recognized (Kumar et al. 2011). In this sense, Locher and Scanlon (2012) consider, as a starting point for the development of sustainable hydroelectric energy, the integrated management of hydraulic resources focused on the knowledge and rationalization of the use of resources of the basins, in addition to planning for its development in order to reduce its impact. In this way, hydroelectric sustainability is made up of three fundamental components:

- The long-term viability of the hydroelectric project.
- The project's contribution to sustainable development.
- The integration of the different perspectives of sustainability (environmental, social, economic, financial, technical and integrity) (Locher and Scanlon 2012).

1.1 Hydroelectric Sustainability

To fulfil the fundamental components of hydroelectric sustainability, in the 1990s the attention was directed to sustainability aspects related to dam construction and hydroelectric power generation. At the international level, one of the first initiatives to define the problem and the mitigation measures to face it, was the implementation of the agreement on hydroelectric technologies issued by the International Energy Agency (IEA). In this sense, between 1998 and 2000, another important contribution was offered by the World Commission on Dams (WCD), in addition to the monitoring and progress of the projects of the United Nations Environment Program. In the second decade of the 21st century, the International Association of Hydroelectricity (IHA) offered the most far-reaching and influential instruments that addressed sustainability in the hydropower sector (Grisales and Murillo 2013.4).

The Hydroelectric Sustainability Assessment Protocol (IHA 2010) benefited, substantially, from many initiatives beyond the WCD, which occurred in the planning of projects and the performance of hydroelectric sustainability. Such initiatives include the Equator Principles, the international performance standards of the International Finance Corporation (IFC), the multinational security policies of development banks (World Bank and Asian Development Bank), the Global Reporting Initiative (GRI), Sustainable Investment Evaluation Tools (for example, Dow Jones Sustainability Index, FTSE4Good), best practices in the hydroelectric sector, as well as business experiences with annual evaluation approaches and sustainability reporting. The Protocol also incorporates the latest experience in integrity procedures at the national, sectoral, institutional and project management levels.

The IHA Protocol contains the sustainability guidelines with the economic, social and environmental dimensions for new hydroelectric projects, and the integrity processes and technical issues relevant for the operation of hydroelectric plants. Sustainability evaluations of hydroelectric projects are carried out through rigorous analysis of objective evidence, carefully looking for key deficiencies, it is a process based on a deep dialogue with the project manager and the consultation to relevant stakeholders. In addition, during its planning stage, a multi-criteria analysis is required in the areas of demonstrated need and strategic adjustment; as well as the location and design for decision making. The IHA Protocol is not intended to certify energy companies in relation to sustainability criteria, but to serve as a guide for governments and planners in the decision-making process, regarding the viability of a project in the dimensions of sustainability, together with a methodology to measure risk and report on the strengths and negative impacts of the project, allowing a quantitative and qualitative comparison of hydroelectric projects (Grisales and Murillo 2013).

1.2 Hydroelectricity in Mexico

In Mexico there are changes in its gross electricity generation park, from 2002 to 2017 it went from 200,362,388 MWh to 257,416,682 MWh, which represents a 22.16% increase. There were changes in the participation of effective capacity by electrical technology; combined cycle technology increased its participation from 21.99% to 49.88%, which represents 27.89%. Other technologies that also increased their participation were carbo-electric, wind, internal combustion and photovoltaic. On the contrary, steam technology reduced its participation from 39.58% to 16.14%, which represents a decrease of −23.44%; the other technologies that also diminished their participation were dual, turbo-gas, hydroelectric, nuclear power and geothermal. Regarding hydroelectricity, despite increasing its generation by 9.6% in this period, its participation decreased to 11.68% (SIE [computer software] 2010).

In Mexico, the long-term energy sector planning instrument is the Estrategia Nacional de Energía (ENE) (National Energy Strategy), which is renewed every year to a fifteen-year horizon (SENER 2014). With the ENE and the Plan Nacional de Desarrollo (National Development Plan) (SENER 2019), the Programa Sectorial de Energía (Energy Sector Program) (PROSENER 2013) is built. With the above documents, the Programa Especial para el Aprovechamiento de Energías Renovables (Special Program for the Use of Renewable Energies) (PEAER 2014) is carried out. From the Ley para el Aprovechamiento de Energías Renovables (Law for the Use of Renewable Energies), the Financiamiento de la Transición Energética (Financing of the Energy Transition) (LAERFTE 2013) and the Ley General de Cambio Climático (General Law on Climate Change) (LGCC 2018), the percentages of generation with fossil sources are specified of 65% for 2024, 60% by 2035, and 50% by 2050. In turn, the LGCC (2018), establishes an indicator of 35% of the contribution of clean technologies for the year 2024.

The challenges these objectives establish to integrate renewable energies into the Sistema Eléctrico Nacional (SEN) (National Electric System) will be enormous. The Prospectiva de Energías Renovables (Renewable Energies Prospective) (SENER 2013) forecasts low and high alternatives, with capacities of 31,147 MW and 38,146 MW for 2026, on the other hand, the Prospectiva del Sector Eléctrico (Electric Sector Prospect) (SENER 2014) indicates alternatives of 26,742 MW and 54,892 MW in the same period. In both prospects, large hydroelectric and wind power plants dominate. The increase in the contribution of renewable energies to the SEN will cause potential changes in energy planning, will modify the growth criteria of the SEN, the operation and control rules of the SEN will be modified, and it will require the creation of new technical profiles and experts (Huacuz 2013).

Taking into account the criterion of the lowest cost in the planning scenario, the growth of the SEN and the compliance with the legal framework of the LAERFTE and the LCC are impossible, therefore, the integration of a set of projects that add environmental costs and of carbon emissions derived from its objectives. But if the criterion of the lowest cost is met in planning, the growth of the SEN and compliance with the legal framework of the LAERFTE and the LCC are impossible, the integration of a set of projects that include in their objectives environmental costs and of carbon emissions, will be promoted. This situation is also reflected in the Coordinación de Proyectos Hidroeléctricos (CPH) (Coordination of Hydroelectric Projects), so its problems resemble those presented here.

1.2.1 CFE's Hydroelectric Projects

The Coordinación de Proyectos Hidroeléctricos (CPH) (Coordination of Hydroelectric Projects), belonging to the Comisión Federal de Electricidad (CFE) (Federal Electricity Commission), is responsible for planning, designing, supervising and building the civil works and electromechanical assembly of the hydroelectric infrastructure required to cover the request for hydroelectric energy in Mexico (CPH 2014). Decisions to integrate new projects to grow in the SEN are made many years in advance, the periods of time from the planning decision until their entry into generation are extensive. It takes around four to seven years to analyse the alternatives to make the decision to build a new hydroelectric plant, so, the provisions lead to a long-lasting economic consequence, since the useful life of hydroelectric plants can be longer than 30 years (CFE 2011).

For CFE (2014), in recent years, there are delays in electricity generation projects, caused primarily by the increase in the processing times of land use licenses and environmental permits; on the other hand, there are the clarifications with the indigenous communities affected, restrictions on authorizing investment, deferrals in the execution of construction work, social complications with municipal authorities and land owners. According to the Inter-American Association for the Defence of the Environment

(AIDA), in Latin America in 2000, Mexico ranked second in the number of displaced people due to the construction of large dams, with more than 170 thousand displaced people, and third in terms of areas flooded by the reservoirs that make up the dams with 307,000 ha. (Kopas & Puentes 2009).

Based on the changes in the country's economic growth hope, which directly incurs the appreciation of electricity demand, as part of the planning process each year the portfolio of projects that will begin to generate energy is updated. When observing the deferred hydroelectric projects in the Capacity Requirement Programs (PRC) from 2004 to 2013, it is observed that, of the nineteen projects programmed in this period, only three were built, El Cajon Hydroelectric Projects in 2007, La Yesca in 2013 and the construction process of Chicoasén II began in 2015, being the last project that CFE built.

1.3 Research Question

Based on this problem, it was proposed to harmonize the planning functions, their social and environmental impacts, seeking to complement the evaluation of hydroelectric projects, to stop being solely technical-economic to a technical-economic-social-environmental and integrity evaluation, to improve decision-making from a sustainable development vision. From this approach, the research question was generated: To what extent does the application of a methodology to assess the sustainability of hydroelectric projects that incorporate multi-criteria decision-making methods in their planning impact the selection of sustainable hydroelectric projects in Mexico?

To address this question, the theoretical framework is presented, as well as the research techniques and instruments that were used to measure the conditions of the criteria and policies established by the CFE regarding sustainable development in the planning stage of hydroelectric projects. From here, progress was made in obtaining more specific data that was required as the search progressed. From the results obtained, it was possible to build a relationship, on the one hand, between the AHP multicriteria decision-making method and that of GP, later on, using the IHA protocol to determine sustainability profiles with a triangulation by methods that used the analysis AHP sensitivity.

2 Methodological Framework

In this work a mixed methodology is added that enriches the scope of the reference context and is part of a larger study. Mixed methods describe research that employs mixed or multiple strategies that answer research questions and test hypotheses (Driessnack et al. 2007).

After defining the type of research, the population of hydroelectric projects where the sample for qualitative research was obtained is identified, as well as the identification of the IHA sustainability criteria for conducting quantitative research. It is structured in a mixed method design. Based on the above, a research scheme was developed (see Fig. 1).

The mixed design used includes two stages, in chronological order, defined in Phase I and Phase II. In the first, it corresponds to the quantitative approach, which was developed first. Phase II is a qualitative approach, for its design and development, it is based on the first. They are described below. As for the type classification and mixed research design, it corresponds to a mixed method design, with dominant status and sequential order when → WHAT. Their strategy is explicit sequential, with quantitative-qualitative sequence, combining the data from the analysis and emphasizing the explanation and interpretation of the relationships (Creswell 2011).

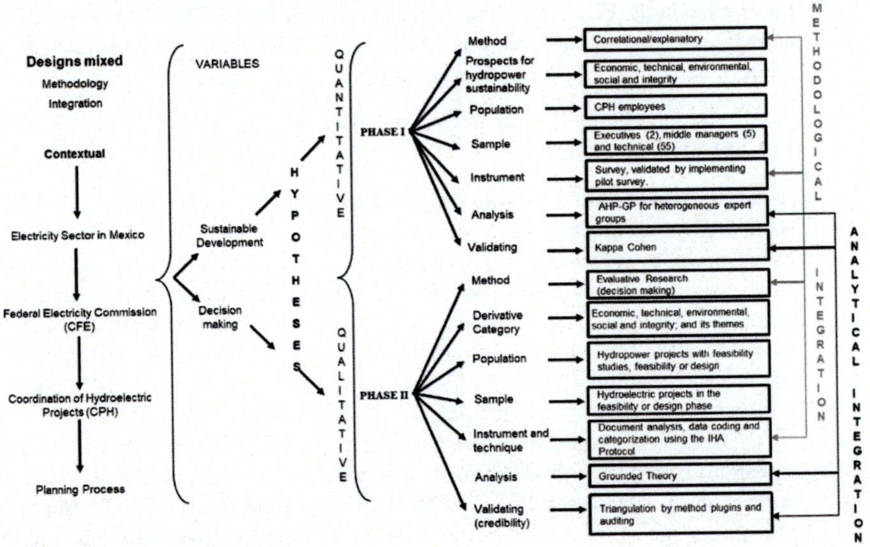

Fig. 1. Methodological research scheme with mixed design. Source: Own elaboration based on Johnson and Onwuegbuzie 2004 and Pereira 2010.

2.1 Phase I: Correlational Method

The correlational method was used, for Hernández et al. (2010), confirm that the correlational studies are intended to know the correlation or degree of association that two or more categories, variables or concepts present in a context. With the premise of establishing the values of the criteria of sustainable development to make decisions in the planning process of hydroelectric projects, for this the experience and knowledge of a group of experts from the CPH is used, through the application of a survey applying the AHP-GP methods that get the opinion of heterogeneous experts in the planning of hydroelectric projects.

2.1.1 Hierarchical Analysis Process (AHP)

To face problems with multiple criteria and a certain number of alternatives, Saaty (1980) presented the AHP, this method allows to graphically divide and organize the problem and order the hierarchies. Through paired pairwise comparisons, it establishes the hierarchy and influence of the fractions that make up the problem, exposes contrasting reflections of value, using the fundamental scale, quantitative and qualitative criteria. Likewise, it agrees to contrast the consistency of value reflections and contributes to improving decision-making (Saaty 1980). It uses fundamental numerical scales for paired comparisons (Table 1) that reflect thoughts, judgments, and intuitions (Casañ 2013).

Table 1. Saaty's fundamental scale

Numerical scale	Verbal scale	Explanation
1	Same	Two activities contribute equally to the objective
3	Moderately	Experience and judgement slightly favour one activity over another
5	Strong importance	Experience and judgement strongly favour one activity over another
7	Very strong or demonstrated importance	An activity is favoured very strongly over another; its dominance demonstrated in practice
9	Extreme importance	The evidence favouring one activity over another is of the highest possible order of affirmation
2, 4, 6, 8	Middle terms	Intermediate values, which are used to express preferences

Source: Saaty (1997) cited by Casañ (2013).

The usefulness of AHP is not limited to being used with intangible criteria and alternatives, therefore, it is used to solve multicriteria decision-making problems, which are choice problems where alternatives are evaluated with respect to various criteria that may be qualitative, quantitative or the combination of both. Also, the AHP has the power to make group decisions that allows decision makers to build group welfare functions that do not violate the conditions of management and strengthen negotiations and cognitive learning (Moreno and Vargas 2018).

2.1.1.1. Adding Preferences

AHP can be used individually or in groups, with the participation of expert groups, obtaining a result that will be the solution that integrates the opinion of those. According to Aznar and Guijarro (2012), the aggregation of opinion can be calculated for each of the experts in two ways:

- Homogeneous experts: they are experts who form a group that have similarities between them, such as academic training, the workplace, and the objective of their work.

- Heterogeneous experts: they are experts who form a group that has a different point of view. This is due to differences in their academic training, due to political situations, among other circumstances.

2.1.1.2. Aggregation of Preferences for Heterogeneous Experts
When the AHP method meets with groups of experts with a discrepancy between their weightings and what they aspire to, find a solution that is not ideal, but rather find a range of solutions where it is possible to assess the different sensitivities of a problem, in this situation will use the goal programming method (Forman and Peniwati 1998).

2.1.2 Goal Programming

Goal programming was published by Charnes et al. (1955), as an extension of linear programming that encompasses multiple criteria. It is used when there is the conflict of achieving multiple objectives in the same period, to achieve it, the decision maker chooses preselected goals by proximity, reducing their deviation. The GP tries to find compromise solutions that, although they do not meet all the goals, do allow achieving certain levels of satisfaction. GP has been combined with other methods, such as AHP (Aznar and Estruch, Valuation of environmental assets 2015).

2.1.2.1. Extended GP for a Diverse Group of Experts
In some evaluations, different groups of evaluators who have incompatible interests may coincide. For example: in the environmental evaluation, as a natural space, different values are found according to the training of the participating social evaluators (administrators, technicians, businessmen, ecologists). In these cases, they were studied by Linares and Romero (2002), Reyna and Cardells (1999), the paired matrices and the eigenvectors of each one of the experts who concentrate by similarity of the hierarchy and by group to obtain a vector were specified. added using the geometric mean. When the different opinions incorporated in the aggregated eigenvectors were combined, another aggregation of the groups was carried out, this time using the Extended Goal Programming, with which it was possible to know the solution closest to the opinion issued by the different experts, as well as establishing the results closer to the opinion of each group of experts. Thus:

- To add predilections of homogeneous groups, the geometric mean is used.
- To add predilections of heterogeneous groups, with different opinions, the use of Programming by Goals (GP) is used.

According to Aull-Hyde et al. (2006), Aznar and Estruch (2007), Ramanathan and Ganesh (1994), there are other preference aggregation methods proposed by other authors, such as the arithmetic mean or the geometric mean (Aczél and Saaty 1983; Aznar and Guijarro 2012; Gass and Rapcsák 1998). Here, the proposed solution was chosen using GP since, in this way, the final solution is given by the median, which allows it to be less affected by anomalous data or assessments whose existence can occur relatively frequently in these cases.

2.1.3 Application of the AHP - GP Method

In the recent literature, references were found that propose the combination of the AHP and GP methods for solving problems related to multiple decisions, for example, Gass and Rapcsák (1998) to carry out an expert group decision synthesis based on the principle of Bridgman; Schniederjans et al. (1995) for a buyer's selection of a home with a number of qualitative and quantitative factors; Kim et al. (1999) for the selection of optimal nuclear fuel cycle scenarios in Korea; Badri (2001) to help select the quality control instrument for collecting customer information; Yurdakul (2004) to select integrated technology in computer manufacturing from competitors; Perçin (2006) for the selection of literature providers; Aznar and Estruch (2007) to find the economic value of an environmental component and its partial values that make it up; Mendoza et al. (2008) to select suppliers with a variety of qualitative and quantitative factors; Erdem and Göçen (2012) to improve supplier assessment and supply chain order allocation decisions; Kozioł (2014) for the valuation of real estate with limited information in Warsaw; Kambiz et al. (2016) developed a model to design and explain cost management in the home appliance industry in Iran; Hamurcu et al. (2017) to select the planning of railway projects in Istanbul; Özan et al. (2017) to select the maintenance strategy in hydroelectric power plants that have great importance in the global energy mix and in Turkey; and Wichapa and Khokhajaikiat (2017) proposes a model to select new suitable places for the elimination of infectious waste.

As can be seen in the previous review of different papers allows to identify that there is no direct application of AHP-GP for hydroelectric sustainability, hence the importance of applying it from an individual or collective perspective. In the latter, a heterogeneous group of experts can get involved, obtaining a useful solution and, consequently, the opinion of all the experts (Aznar and Estruch 2015).

2.1.4 Kappa Index

Since its introduction in 1960, the Kappa coefficient has become the statistician for evaluating agreements between judges as was obtained by Cohen (1960), especially, the probability of the agreement between them. This probability was credited as Cohen's Kappa coefficient, which is a statistic of agreement between judges that corrects chance. For Viera and Garrett (2005), the Kappa coefficient is a scale of -1 to 1, where one indicates a perfect agreement, 0 is exactly what would be expected by chance, and negative values indicate that the agreement obtained will be less than that obtained randomly, that is, a potential systematic disagreement between observers in order to facilitate their interpretation (Landis and Koch 1977).

2.2 Phase II: Documentary Analysis

In this work the Grounded Theory method was used (Glaser and Strauss 1967), it is a method of qualitative analysis of semi-structured interviews in depth. This process resides in the codification of the empirical data by means of words derived from the incidents, occurrences or events that the interviewees express; later they are grouped into categories, concepts or constructs, and from here the differences and similarities that exist between one category and another are identified (Glaser and Strauss 1967).

The IHA protocol in preparation is used for the coding and data category process. Through matrices, the information that facilitates the evaluation of objective evidence will be displayed to determine hydroelectric sustainability profiles of the projects. For the documentary analysis of Phase II, the evaluation of the sustainability of hydroelectric projects described in the IHA Protocol (2010), was used.

The hydroelectric sustainability assessment is made through a strict analysis of objective evidence, scrupulously reviewing key flaws, it is a procedure that is based on an in-depth dialogue with the project manager and its relevant stakeholders.

After exposing the approach, description of the problem, as well as the type and design of the study, the selection of hydroelectric projects with sustainability criteria is presented below, using the AHP and GP methods, in non-homogeneous groups of experts, in order to calculate the priorities of criteria, sub-criteria and alternatives.

3 Materials and Methods

The research carried out is expected to help decision-makers in selecting hydroelectric projects, so a survey was used to determine the weightings of the five criteria of the hierarchical model of hydroelectric sustainability (Fig. 2).

In this case, twenty-four sub-criteria (topics) were identified in the preparation stage of the IHA Protocol (IHA 2010), in the perspective of integrity, the issue of evaluation and management of social and environmental impact, it was considered necessary to separate the part social and environmental (Fig. 2), since there are differences in the management of these issues in hydroelectric projects.

The alternatives identified are the Chicoasén II, Nuevo Guerrero and Las Cruces hydroelectric projects. Since these projects are the ones that have the most progress with respect to the Preparation stage of the IHA protocol that concludes with the award of the construction contract. One of the most important stages of the AHP method is to structure the hierarchy of the problem, which is made up of the objective, the criteria, the sub criteria, and the alternatives.

3.1 Measuring Instrument

The survey was the instrument used to obtain the paired matrices, required in the AHP method described; This allowed determining the importance of each hydroelectric sustainability criterion. The questionnaire was validated by six experts, three experts in hydraulics, two in methodology and one in multi-criteria methods.

3.1.1 Collective

In the AHP, the knowledge, experience and importance of the group surveyed are of paramount importance, rather than the number of people to be interviewed. The AHP examines experts regarding the object of study. Quantitative research accepts sampling for convenience (Schwab 1995), considering that they should look for the most representative and convenient units for the study (Grande and Abascal 2014). From August to October, 2018, the survey was applied in a personalized way and by email, with an average duration of twenty minutes per respondent. Due to the complexity of hydroelectric

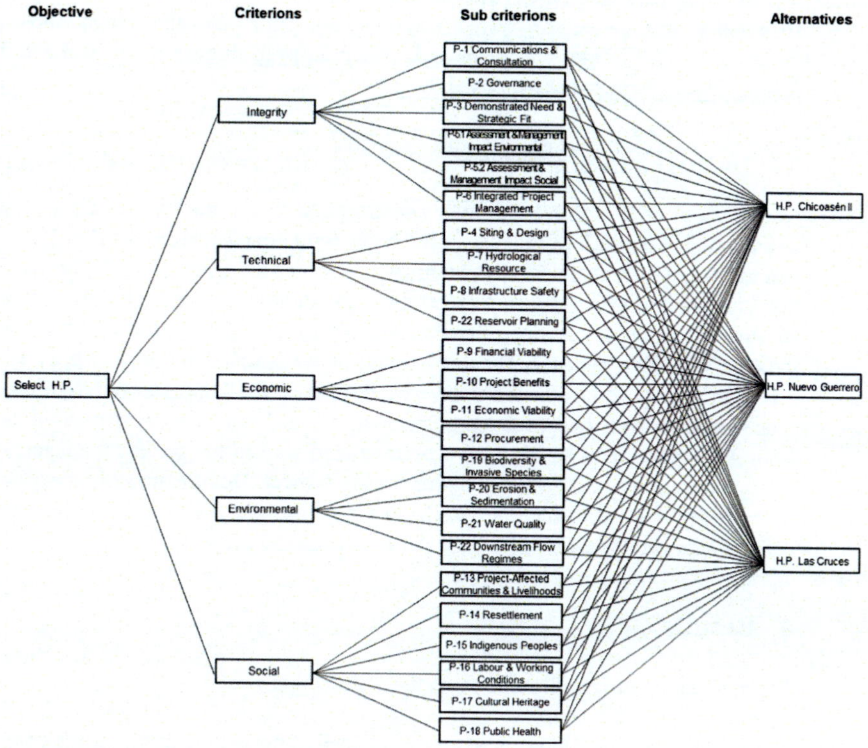

Fig. 2. Hydroelectric sustainability hierarchies tree for the preparation stage. Source: Own elaboration based on the results obtained with Web HIPRE 2007.

projects and their characteristics and political and social implications, it was considered appropriate to carry out the survey of a group of experts that covered the different areas involved in the planning of hydroelectric projects. So, the people surveyed were 64, grouped according to their specialty.

The group of experts was made up of various professionals such as: anthropologists, biologists, sociologists, environmental engineers, industrial engineers, civil engineers, hydraulics engineers, electrical engineers, electromechanical engineers, economists, business administrators; causing a diversity of different criteria and points of view to be presented on the hierarchy of each criterion and sub criteria, including within each specialty. Of the 64 experts, 41 of them have studies at bachelor's level, 21 at master's level and two have doctorates. His experience in planning hydroelectric projects ranges from 4.59 to 26.67 years, with an average experience of 10.70 years. 22% are women and 78% are men.

Thus, experts in the planning process were considered responsible for decision-making and those who, directly and indirectly, influence it. For this, there are five groups of experts: economic 6; technicians 38; environmental 6; social 7, and integrity 7.

Finally, the eigenvectors were obtained by using the geometric mean. Subsequently, the extended GP model was proposed, which allowed finding the useful solution and identifying the group that proposed it.

3.2 Documentary Analysis

Comparisons of alternatives for hydroelectric projects required a documentary analysis, as indicated in phase II of the research scheme with mixed design (see Fig. 1), in order to determine the hydroelectric sustainability profiles of the Chicoasén II, Nuevo Guerrero and Las Cruces hydroelectric projects. For the documentary analysis, the methodology for the evaluation of sustainability issues was used; in each hydroelectric sustainability issue, an objective review of the documentation was carried out, in which, in a first stay, the basic good practices were reviewed and if they were not presented Significant fouls proceeded to review best practices. The documentary analysis was applied in the period from August to October 2018, in a personalized way to the project managers, with an average duration of 30 days per hydroelectric project. Documentary was applied in the period from August to October 2018, in a personalized way to the project managers, with an average duration of 30 days per hydroelectric project.

4 Results

The analysis of the results addresses the following aspects:

- Results of the test. Determine the consistency of the results of the paired comparisons using the Saaty scale (Table 2).
- Form groups by specialty and hierarchical level to identify the group that proposes the best solution.

It is observed in Table 2 that, with the exception of respondents 24 and 58 (in italics and marked with an *), who obtained a CR greater than 10% and were inconsistent, the others reached a CR less than or equal to 10%. This left a total of 62 consistent surveys (97%) that were grouped according to the specialty of the respondents in: economic, technical, environmental, social and integrity.

4.1 Results by Homogeneous Expert Groups by Specialty

Aggregate eigenvectors were calculated for each group of experts using the geometric mean. Eigenvectors were calculated for each group using the geometric mean. The normalized values of all the groups are presented in Table 3.

Table 3 presents the results by group of experts: economic and technical experts give higher priority to the economic perspective and lower priority to that of integrity; environmental experts give higher priority to the integrity perspective and lower priority to technique; social experts give higher priority to the social perspective and lower priority to technique; Finally, integrity experts give higher priority to the economic perspective and lower priority to the environmental one.

Table 2. Radius of consistency of the surveys

Survey	RC	Survey	RC	Survey	RC
1.	0.0000	23.	0.0703	45.	0.0781
2.	0.0000	24.*	0.1739	46.	0.0888
3.	0.0415	25.	0.0787	47.	0.0000
4.	0.0108	26.	0.0572	48.	0.0000
5.	0.0000	27.	0.0984	49.	0.0957
6.	0.0801	28.	0.0420	50.	0.0553
7.	0.0709	29.	0.0504	51.	0.0681
8.	0.0959	30.	0.0745	52.	0.0310
9.	0.0901	31.	0.0734	53.	0.0695
10.	0.0373	32.	0.0918	54.	0.0963
11.	0.0433	33.	0.0977	55.	0.0306
12.	0.0985	34.	0.0358	56.	0.0788
13.	0.0489	35.	0.0831	57.	0.0577
14.	0.0125	36.	0.0240	58.*	0.2202
15.	0.0663	37.	0.0853	59.	0.0000
16.	0.0272	38.	0.0881	60.	0.0922
17.	0.0681	39.	0.0944	61.	0.0353
18.	0.0880	40.	0.0819	62.	0.0986
19.	0.0000	41.	0.0612	63.	0.0258
20.	0.0999	42.	0.0000	64.	0.0689
21.	0.0537	43.	0.0749		
22	0.0621	44.	0.0404		

Source: Own elaboration.

When comparing the results of the five groups of experts using a concordance matrix (Table 3), a Kappa Index of 92.82% reliability is obtained, so the degree of concordance is almost perfect.

4.1.1 GP Application Extended to Heterogeneous Expert Groups by Specialty

The solution closest to the opinion issued by the expert groups by specialty was sought and the solution closest to the opinion of each group was determined. Here the extended GP model was also proposed (5), three solutions were obtained as a function of λ (Table 4) using the LINGO software (LINDO-Systems 2019).

Table 3. Eigenvectors by group of experts

Experts by specialty	Eigenvectors				
	Economic	Technical	Environmental	Social	Integrity
Economic	0.4498	0.1533	0.1517	0.1486	0.0967
Technical	0.2959	0.1987	0.1684	0.2198	0.1172
Environmental	0.1910	0.1447	0.2063	0.2076	0.2504
Social	0.1390	0.0991	0.2297	0.2685	0.2638
Integrity	0.2880	0.2101	0.1451	0.1929	0.1638

Source: Own elaboration.

The solution was the one that presents the λ values between 0.2 to 1, in this interval a minimum Z value of 2.2460 was achieved. This solution was the closest to the opinions issued by the integrity expert group (D5). When considering the best solution, the weighting of the hydroelectric sustainability prospects is included in Table 5.

The weights in Table 8 indicate that the best positioned perspective was the economic one, since it obtained a priority of 29.038%. The technical perspective was placed next with a priority of 21.039%. Finally, the perspective at the lowest level was environmental, with a priority of 14,365%.

4.1.2 Analysis of Solutions

The results of the 62 surveys were organized into five groups by specialty. Using the geometric mean and extended GP models, a useful solution was found for the group. Next, the results obtained from the groups are compared, emphasizing the solution that is closest to the opinion of each of the expert groups (Table 6).

When comparing the useful solution with the results of the five groups, it is observed that the opinion closest to the solution is that of the group of integrity experts that obtained a difference of 0.00525. Therefore, it is stated that the survey applied to this group presented disparate opinions. This allowed validating the use of the AHP and extended GP multicriteria methods to determine weights of the hydroelectric sustainability criteria by means of the survey of heterogeneous groups of experts.

4.1.3 Sub-criteria Weighting

After the results of the useful solution (see Table 5), surveys were applied to five experts in each of the hydroelectric sustainability perspectives, who additionally had knowledge of the IHA Protocol in order to determine the weightings of the Social, Technical sub-criteria. Economic, Environmental and Integrity. The Fig. 3 shows the global and local weights of the sub-criteria.

4.2 Documentary Analysis

By means of graphs, the information will be displayed to facilitate the evaluation of the objective evidence that determines the sustainability profile (see Fig. 4).

Table 4. Landa-based solutions with LINGO

λ	W1	W2	W3	W4	W5	Z	D	D1	D2	D3	D4	D5
0	0.29440	0.19845	0.13345	0.18485	0.15215	2.3040	0.15440	0.15540	0.03495	0.10340	0.15540	0.01165
0.1	0.29440	0.19330	0.13860	0.19000	0.14700	2.3040	0.15440	0.15440	0.02980	0.10340	0.15540	0.01680
0.2–1	0.28280	0.20490	0.13990	0.18770	0.15860	2.2460	0.16700	0.16700	0.04140	0.09180	0.14380	0.00520

Source: Own elaboration.

Table 5. Final weighting of the prospects for hydroelectric sustainability

Perspective	Weighting	Normalized weighting
Economic	0.2828	0.29038
Technic	0.2049	0.21039
Environmental	0.1399	0.14365
Social	0.1877	0.19273
Integrity	0.1586	0.16285
	0.9739	1.0000

Source: Own elaboration.

Table 6. Comparison of the solution with the results of the expert groups

Values	Solution	Speciality				
		Economic	Technic	Environmental	Social	Integrity
Economic	0.29038	0.44976	0.29586	0.19102	0.13900	0.28801
Technic	0.21039	0.15331	0.19874	0.14474	0.09908	0.21013
Environmental	0.14365	0.15166	0.16840	0.20630	0.22967	0.14512
Social	0.19273	0.14861	0.21978	0.20759	0.26848	0.19294
Integrity	0.16285	0.09667	0.11722	0.25035	0.26378	0.16379
Difference		**0.33477**	**0.11456**	**0.33001**	**0.52538**	**0.00525**

Source: Own elaboration.

For credibility effects in the work by method triangulation, the AHP-GP method (quantitative approach criteria) is complemented with the documentary analysis (qualitative approach sub-criteria), to establish a strong triangulation with the determination of the alternatives in the synthesis of the AHP.

4.2.1 Weighting of Alternatives

In the same way that the paired comparisons between the criteria and the sub criteria were made, the 3 alternatives were compared with respect to the 24 sub criteria of the model. To determine the local and global weights of each of the elements of the hierarchical model, all the weights calculated for each component of the model must be synthesized (see Table 7).

4.3 Model Synthesis

Once the vector resulting from the pairwise comparisons between the criteria has been calculated, Table 8 shows: (i) the pairwise comparisons between the three hydroelectric projects (vector) and (ii) their prioritization for each criterion and sub-criterion analysed.

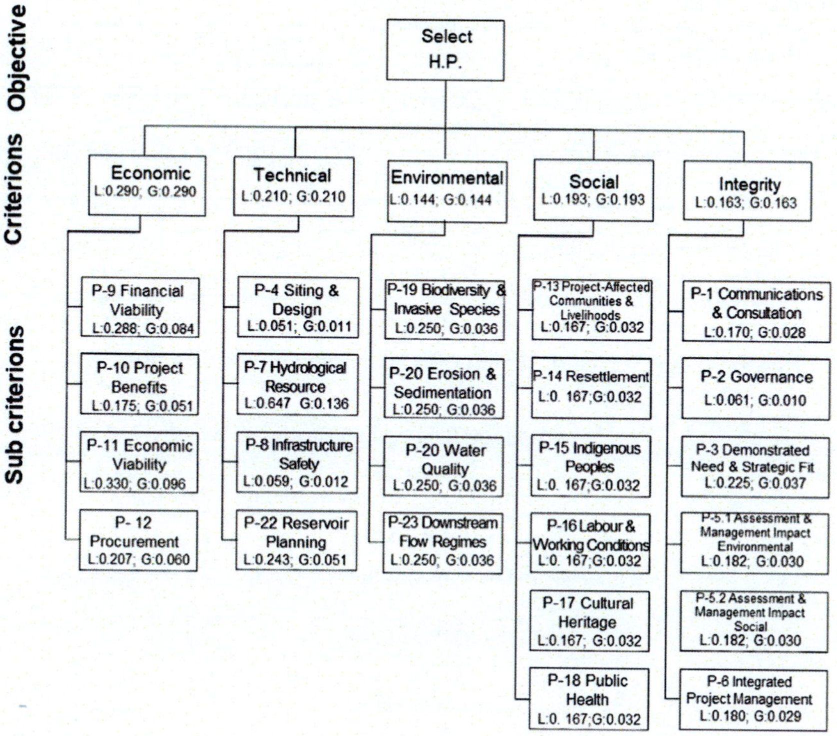

Fig. 3. Global and local weightings of criteria and sub-criteria of the hierarchical model. Source: Own elaboration.

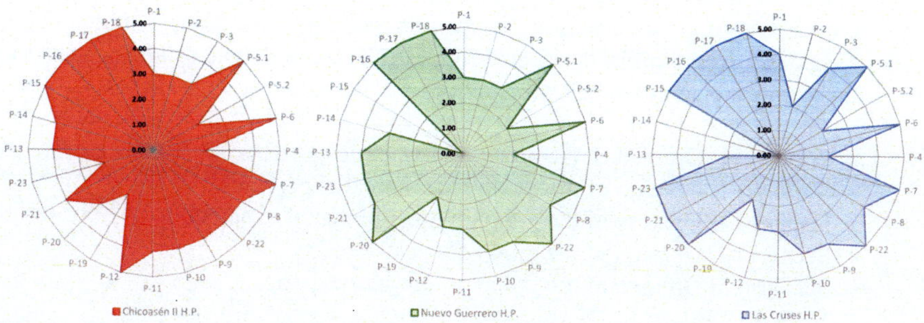

Fig. 4. Hydroelectric sustainability profiles of the hydroelectric projects Chicoasén II, Nuevo Guerrero and Las Cruses. Source: Own elaboration.

To make relative comparisons between equivalent hydroelectric projects, the AHP method sorts the projects from best to lowest, simultaneously considering the relationship of the criteria and the detailed sub-criteria.

Table 7. Local Priorities obtained for each sub criteria of hydroelectric sustainability of the hierarchical model

Alternatives	Integrity sub criteria						Technical sub criteria				Economic sub criteria				Environmental sub criteria				Social sub criteria					
	P-1 Communications & Consultation	P-2 Governance	P-3 Demonstrated Need & Strategic Fit	P-5.1 Assessment & Management Impact	P- Environmental Management Impact	P-6 Integrated Project Management	P-4 Siting & Design	P-7 Hydrological Resource	P-8 Infrastructure Safety	P-22 Reservoir Planning	P-9 Financial Viability	P-10 Project Benefits	P-11 Economic Viability	P-12 Procurement	P-19 Biodiversity & Invasive Species	P-20 Erosion & Sedimentation	P-21 Water Quality	P-22 Reservoir Planning	P-13 Project-Affected Communities & Livelihoods	P-14 Resettlement	P-15 Indigenous Peoples	P-16 Labour & Working Conditions	P-17 Cultural Heritage	P-18 Public Health
Chicoasen II H.P.	0.250	0.250	0.250	0.333	0.333	0.333	0.333	0.333	0.333	0.200	0.333	0.333	0.400	0.600	0.333	0.143	0.250	0.122	0.429	0.493	0.429	0.333	0.333	0.333
Nuevo Guerrero H.P.	0.250	0.500	0.250	0.333	0.333	0.333	0.333	0.333	0.333	0.400	0.333	0.333	0.200	0.200	0.333	0.429	0.250	0.320	0.429	0.311	0.143	0.333	0.333	0.333
Las Cruces H.P.	0.500	0.250	0.500	0.333	0.333	0.333	0.333	0.333	0.333	0.400	0.333	0.333	0.400	0.200	0.333	0.429	0.500	0.558	0.143	0.196	0.429	0.333	0.333	0.333
Weightings	**0.170**	**0.061**	**0.225**	**0.182**	**0.182**	**0.180**	**0.051**	**0.647**	**0.059**	**0.243**	**0.288**	**0.175**	**0.330**	**0.207**	**0.250**	**0.250**	**0.250**	**0.250**	**0.167**	**0.167**	**0.167**	**0.167**	**0.167**	**0.167**

Source: Own elaboration based on the results obtained with Web-HIPRE.

Table 8. Ranking of hydroelectric projects

Alternatives	Global weighting	Ranking
Chicoasén II H.P	0.345	2
Nuevo Guerrero H.P	0.306	3
Las Cruces H.P	0.349	1

Source: Own elaboration based on the results obtained with Web-HIPRE.

4.4 Sensitivity Analysis

The analysis identifies, analyses, and shows the sensitivity of the results, considers the criteria with potential changes that may cause significant changes in the results of the weightings of the projects selected in the AHP method. The following scenarios from the survey are analysed below.

- Scenario 1: Opinion of the group of economic experts.
- Scenario 2: Opinion of the group of technical experts.
- Scenario 3: Opinion of the group of environmental experts.
- Scenario 4: Opinion of the group of social experts.
- Scenario 5: Opinion of the group of integrity experts.

This analysis uses Web-HIPRE computing tools. The analysis makes variations in the criteria weights and shows, numerically and graphically, how the changes affect the rest of the values of the alternatives and their ranking (see Fig. 5).

The results obtained from Fig. 5 show the five scenarios and the modifications in the criteria weights. The alternative of the Las Cruces Hydroelectric Project. Only in the scenario of the group of economic experts the alternative of the Chicoasén II Hydroelectric Project occupies the first place. Finally, the Nuevo Guerrero Hydroelectric Project in all scenarios takes last place.

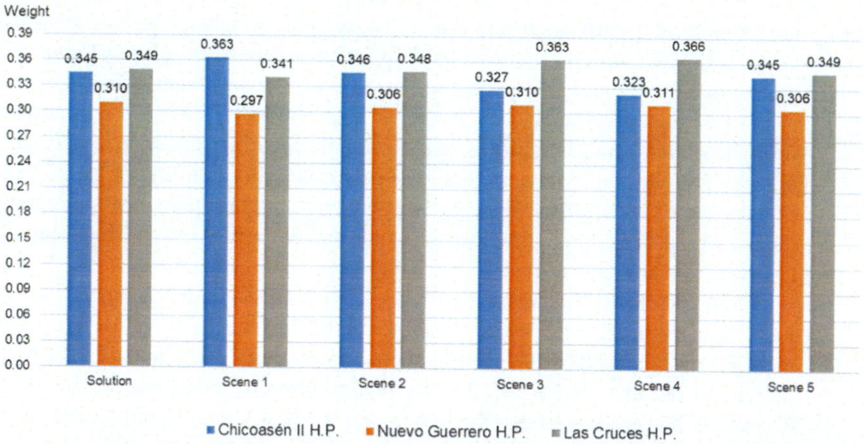

Fig. 5. Sensitivity analysis of the Ranking of alternatives. Source: Own elaboration

The Las Cruces Project is the most appropriate in the evaluation with hydroelectric sustainability criteria. After carrying out an analysis of the selection of hydroelectric projects with sustainability criteria and applying the multi-criteria AHP and GP decision-making methods to heterogeneous groups of experts, the conclusion is reached of selecting the Las Cruces hydroelectric project, as the project best evaluated using hydroelectric sustainability criteria.

5 Conclusions

Experts in hydroelectric project planning are increasingly aware of sustainable development and the balance that projects must maintain. The social part of a hydroelectric project is transcendental to be able to build the new hydroelectric plants. Good management and stakeholder involvement will help improve decision-making to select the best sustainable hydroelectric project.

AHP-GP methods are an approach that manages to obtain the opinion of a heterogeneous group of experts and thus determine the weights of the criteria for hydroelectric sustainability through surveys. The AHP allows the use of rational thinking and logical arguments to select criteria and determine weights through paired comparisons, while the GP allows a useful solution to be obtained from heterogeneous groups by surveying them. In this context, the Kappa index is used to evaluate agreements between judges, and it was used in other multicriteria decision-making studies to verify its accuracy.

In relation to the analysis of results carried out, it was determined that the group of experts in integrity is the one that proposes the closest solution to the opinion issued by each group. Of the weights obtained, the best positioned perspective was the economic one with 29.04%, then followed, in descending order, the technique with 21.04% and finally the perspective with the lowest weighting was the environmental one with 14.36%.

When questioning the heterogeneous group of experts to determine the weights of the criteria of hydroelectric sustainability, it was observed that significant agreements are presented between the groups when assigning the weights, when presenting a Kappa index = 0.9282, which is striking as they are heterogeneous groups of experts, since a greater disagreement was expected for being professionals from different disciplines.

The documentary analysis was prepared using the IHA (2010) evaluation method for hydroelectric sustainability issues, for this purpose an objective review of the documentation required by this method was performed. For the categorization and coding of data, the protocol of the IHA (2010) was used in its preparation stage; and through the use of matrices, the hydroelectric sustainability profiles of the Chicoasén II, Nuevo Guerrero and Las Cruces hydroelectric projects were determined.

From the sensitivity analysis carried out, five scenarios were obtained from which the Las Cruces Hydroelectric Project remained at the first site in four of the five scenarios. In the scenario proposed by the group of economic experts, the Chicoasén II hydroelectric project ranked first, and finally the Nuevo Guerrero project ranked last in all scenarios. Therefore, it is confirmed that the selection of the Las Cruces hydroelectric project was the best alternative.

Regarding the theoretical implications, it was found that the documentary evaluation of the IHA protocol supported by the AHP-GP method, should be considered as a mixed research method as indicated by Johnson and Onwuegbuzie (2004) and Pereira (2010).

The research provides support to those seeking to assess the sustainability of hydroelectric projects using the IHA protocol and the combination of multi-criteria decision-making techniques used to obtain the weights of the hydroelectric sustainability criteria. However, future research should focus on comparing our results with the results of other expert groups in different contexts of hydroelectric projects.

Taking into account the above, as possible future work, the use of a method integrated by AHP-GP is recommended in the analysis of decision-making for large-scale projects, its application in other energy scenarios is recommended, such as the following: compare electrical projects from different energy sources and with energy sustainability criteria; and define public policies to generate renewable energy and energy security. Likewise, it is proposed to include for future research in the expert groups those responsible for establishing public policies and energy planning, as well as academic experts in social and environmental issues. Finally, deepen the documentary analysis of the social problem in hydroelectric projects in Mexico.

References

Aczél, J., Saaty, T.: Procedures for synthesizing ratio judgements. J. Math. Psychol. **27**(1), 93–102 (1983). https://doi.org/10.1016/0022-2496(83)90028-7

Arancibia, S., Contreras, E., Mella, S., Torres, P.: Evaluación Multicriterio: aplicación para la formulación de proyectos de infraestructura deportiva: Memoria de Ingeniero Civil Industrial Universidad de Chile, Santiago (2003)

Atwi, B., Arrojo, P.: Impacto ambiental de las grandes presas en cursos bajos, deltas y plataformas litorales: el caso de Aswan. Actas del Segundo Congreso Ibérico sobre Gestión y Planificación de Aguas. Recuperado el 28 de enero de 2020, de (2000). http://ocw.um.es/ciencias/ecologia/lectura-obligatoria-1/arrojo-atwi.pdf

Aull-Hyde, R., Erdogan, S., Duke, J.: An experiment on the consistency of aggregated comparison matrices in AHP. Eur. J. Oper. Res. **171**(1), 290–295 (2006). https://doi.org/10.1016/j.ejor.2004.06.037

Ávila, R.: El AHP (Proceso Analítico Jerárquico) y su aplicación para determinar los usos de las tierras. El caso de Brasil. Informe técnico. Proyecto Regional Información sobre Tierras y Aguas para un Desarrollo Agrícola Sostenible (Proyecto Gcp/Rla/126/Jpn), Santiago, Chile (2000)

Aznar, J., Estruch, V.: Valoración de activos ambientales mediante métodos multicriterio. Aplicación a la valoración del Parque Natural del Alto Tajo. Economía agraria y recursos naturales **7**(13), 107–126 (2007)

Aznar, J., Estruch, V.: Valoración de activos ambientales. Editorial Universitat Politècnica de València, Valencia (2015)

Aznar, J., Guijarro, F.: Nuevos métodos de valoración: Modelos Multicriterio. Editorial Universitat Politècnica de València, Valencia (2012)

Badri, M.: A combined AHP–GP model for quality control systems. Int. J. Prod. Econ. **72**(1), 27–40 (2001). https://doi.org/10.1016/S0925-5273(00)00077-3

Brunelli, M., Canal, L., Fedrizzi, M.: Inconsistency indices for pairwise comparison matrices: a numerical study. Ann. Oper. Res. **211**(1), 493–509 (2013). https://doi.org/10.1007/s10479-013-1329-0

Casañ, A.: La decisión multicriterio: aplicación en la selección de ofertas competitivas en edificación. Univerdidad Politécnica de Cataluya, Cataluya (2013)

Castillo, L., Dorao, C.: Decision-making in the oil and gas projects based on game theory: conceptual process design. Energy Convers. Manage. **66**, 48–55 (2013). https://doi.org/10.1016/j.enconman.2012.09.029

CFE: Programa de Obras e Inversiones del Sector Eléctrico 2012–2026. Comisión Federal de Electricidad, México (2011)

CFE: Programa de Obras e Inversiones del Sector Eléctrico: POISE 2014–2028. Comisión Federal de Electricidad. Subdirección de Programación, México (2014)

Charnes, A., Cooper, W., Ferguson, R.: Optimal estimation of executive compensation by linear programming. Manage. Sci. **1**(2), 138–151 (1955). https://doi.org/10.1287/mnsc.1.2.138

CMR: Represas y desarollo: Un nuevo marco para la toma de decisiones: El raporte final de la comision mundial de represas. Earthscan Publications Ltd. Ed., London (2000)

Cohen, J.: A coefficient of agreement for nominal scales. Educ. Psychol. Measur. **20**(1), 37–46 (1960)

CPH: LM 5000 Manual del Sistema de Gestión Integrado (Revisión 8 ed.). Coordinación de Proyectos Hidroeléctricos, México (2014)

Creswell, J.: Controversies in mixed methods research: state of the art. In: Denzin, N., Lincoln, Y. (eds.) The SAGE Handbook of Qualitative Research, pp. 269–284. SAGE, London (2011)

Driessnack, M., Sousa, V., Mendes, I.: Revisión de los diseños de investigación relevantes para la enfermería: parte 3: métodos mixtos y múltiples. Rev Latino-am Enfermagem **15**(3), 502–507 (2007)

Erdem, A., Göçen, E.: Development of a decision support system for supplier evaluation and order allocation. Exp. Syst. Appl. **39**(5), 4927–4937 (2012)

Forman, E., Peniwati, K.: Aggregating individual judgments and priorities with the analytic hierarchy process. Eur. J. Oper. Res. **108**(1), 165–169 (1998). https://doi.org/10.1016/S0377-2217(97)00244-0

Gass, S., Rapcsák, T.: A note on synthesizing group decisions. Decis. Support Syst. **22**(1), 59–63 (1998). https://doi.org/10.1016/S0167-9236(96)00061-9

Glaser, B., Strauss, A.: The Discovery of Grounded Theory: Strategies for Qualitative Research. Aldine, Chicago (1967)

Grande, I., Abascal, E.: Fundamentos y Técnicas de Investigación Comercial. ESIC Editorial, Madrid (2014)

Grisales, E., Murillo, J.: El mercado de bonos de carbono y su aplicación para proyectos hidroeléctricos. Revista CINTEX, pp. 131–143 (2013)

Hamurcu, M., Alağaş, H., Eren, T.: Selection of rail system projects with analytic hierarchy process and goal programming. Sigma **8**(2), 291–302 (2017)

Hamzeh, S., Mokarram, M., Alavipanah, S.: Combination of Fuzzy and AHP methods to assess land suitability for barley: case Study of semi arid lands in the southwest of Iran. Desert **19**(2), 173–181 (2014)

Hernández, Fernández, C., Baptista, P.: Metodología de la Investigación. Mc Graw-Hill, México (2010)

Huacuz, J.: El inventario de las energías renovables en el marco de la transición energética de México. Boletín IIE **1**, 92–100 (2013)

Huck, S.: Reading Statistics and Research. Pearson Education Limited, New York (2013)

IEA: World Energy Outlook 2018. International Energy Agency, Paris, France (2018)

IHA: Protocolo de Evaluación de la Sostenibilidad de la Hidroelectricidad. Asociación Internacional de la Energía Hidroeléctrica, London (2010)

Johnson, R., Onwuegbuzie, A.: Mixed methods research: a research paradigm whose time has come. Educ. Res. **33**(7), 14–26 (2004)

Kambiz, S., Saeed, Y., Mohsen, K.: Mathematical model designing to achieve cost management in value chain with combinational approach of AHP & GP: case study: home appliance industries. Sch. J. **4**(1), 30–51 (2016)

Kim, P., Lee, K., Lee, B.: Selection of an optimal nuclear fuel cycle scenario by goal programming and the analytic hierarchy process. Ann. Nucl. Energy **26**(5), 449–460 (1999)

Kopas, J., Puentes, A.: Grandes Represas en América: ¿Peor el remedio que la enfermedad? Principales consecuencias ambientales y en los derechos humanos y posibles alternativas. Asociación Interamericana para la Defensa del Ambiente (AIDA), Bogotá (2009)

Kozioł, D.: The use of combined multicriteria method for the valuation of real estate. Studia Ekonomiczne **5**(71), 219–235 (2014)

Kumar, A., Schei, T.T., Ahenkorah, A., Rodriguez, R.: Hydropower: IPCC special report on renewable energy sources and climate change mitigation (2011)

Kusre, B., Baruah, D., Bordoloi, P., Patra, S.: Assessment of hydropower potential using GIS and hydrological modeling technique in Kopili River basin in Assam (India). Appl. Energy **87**(1), 298–309 (2010). https://doi.org/10.1016/j.apenergy.2009.07.019

LAERFTE: Ley para el Aprovechamiento de Energías Renovables y el Financiamiento de la Transición Energética. México: D.O.F. 28-11-2008, Ult. Ref. Pub. D.O.F. 07-06-2013 (2013)

Landis, J., Koch, G.: The measurement of observer agreement for categorical data. Biometrics **33**(1), 159–174 (1977)

Latour, J., Abraira, V., Cabello, J., Sánchez, J.: Las mediciones clínicas en cardiología: validez y errores de medición. Rev. Esp. Cardiol. **50**(2), 117–128 (1997)

LGCC: Ley General de Cambio Climático. México: D.O.F. 06-06-2012, Ult. Ref. Pub. D.O.F. 13-07-2018 (2018)

Linares, P., Romero, C.: Aggregation of preferences in an environmental economics context: a goal-programming approach. Omega **30**(2), 89–95 (2002). https://doi.org/10.1016/S0305-0483(01)00059-7

LINDO-Systems: Products. Obtenido de LINGO Optimization Modeling, 22 de 04 de 2019. https://www.lindo.com/index.php/products/lingo-and-optimization-modeling

LINGO 15 [software de computación]: LINDO Systems INC. (Version API Version 9.0.2120.232) (2015)

Locher, H., Scanlon, A.: Sustainable hydropower-issues and approaches. In: Samadi, H. (ed.) Hydropower – Practice and Aplication, pp. 1–22. InTech, Coatia (2012)

Márquez, H.: Métodos matemáticos de evaluación de factores de riesgo para el Patrimonio Arqueológico: una aplicación Gis del método de jerarquías analíticas de TL Saaty. Spal **8**, 21–37 (1999)

Martorell, S., et al.: RAMS+C informed decision-making with application to multi-objective optimization of technical specifications and maintenance using genetic algorithms. Reliab. Eng. Syst. Saf. **87**(1), 65–75 (2005). https://doi.org/10.1016/j.ress.2004.04.009

Maurtua, D.: Criterios de Selección de Personal mediante el uso del proceso de análisis jerárquico. Aplicación en la selección de personal para la Empresa Exotic Foods SAC (Licenciado en Investigación Operativa). Universidad Nacional Mayor de San Marcos, Lima, Perú (2006)

McCully, P.: Ríos silenciados: ecología y política de las grandes represas. Ediciones Proteger (2004)

Mendoza, A., Santiago, E., Ravindran, A.: A Three-Phase Multicriteria Method to the Supplier Selection Problem: International Journal of Industrial Engineering. Theory, Applications and Practice **15**(2), 195–210 (2008)

Moreno, J., Vargas, L.: Decisión Multicriterio Cognitiva y el Legado del Proceso Analítico Jerárquico. Estudios de Economía Aplicada **36**, 67–80 (2018)

Özcan, E.C., Ünlüsoy, S., Eren, T.: A combined goal programming – AHP approach supported with TOPSIS for maintenance strategy selection in hydroelectric power plants. Renew. Sustain. Energy Rev. **78**, 1410–1423 (2017). https://doi.org/10.1016/j.rser.2017.04.039

Paredes, J.: Toma de decisiones. Proceso Analítico Jerárquico aplicado a la selección de herramienta de educación virtual. FPUNE Sci. **6**(6), 95–102 (2014)

PEAER: Programa Especial para el Aprovechamiento de Energías Renovables. México: D.O.F. 28-04-2014 (2014)

Perçin, S.: An application of the integrated AHP-PGP model in supplier selection. Meas. Bus. Excell. **10**(4), 34–49 (2006). https://doi.org/10.1108/13683040610719263

Pereira, Z.: La mirada de estudiantes de la Universidad Nacional hacia el docente y la docente: sus características y clima de aula. (Tesis para obtener elgrado de Doctor). Univerdidad Estatal a Distancia, San José, Costa Rica (2010)

PROSENER: Programa Sectorial de Energía 2013–2018. México: D.O.F. 13-12-2013 (2013)

Ramanathan, R., Ganesh, L.: Group preference aggregation methods employed in AHP: an evaluation and an intrinsic process for deriving members' weightages. Eur. J. Oper. Res. **79**(2), 249–265 (1994). https://doi.org/10.1016/037

Reyna, S., Cardells, F.: Valoración AHP de los ecosistemas naturales de la Comunidad Valenciana. Revista Valenciana d'Estudis Autonomics **57**(2), 153–179 (1999)

Rojanamon, P., Chaisomphob, T., Bureekul, T.: Application of geographical information system to site selection of small run-of-river hydropower project by considering engineering/economic/environmental criteria and social impact. Renew. Sustain. Energy Rev. **13**(9), 2336–2348 (2009). https://doi.org/10.1016/j.rser.2009.07.003

Saaty, T.: The Analytic Hierarchy Process. McGraw Hill, New York (1980)

Saaty, T.: Toma de decisiones para líderes: el proceso analítico jerárquico en la toma de decisiones en un mundo complejo. RWS Publications, Pittsburgh (1997)

Saaty, T., Vargas, L.: Models, Methods, Concepts & Applications of the Analytic Hierarchy Process. Springer, New York (2012). https://doi.org/10.1007/978-1-4614-3597-6

Salas, P. (s.f.): Recuperado el 17 de 03 de 2020, de Energía hidroeléctrica central Pangue. http://cipres.cec.uchile.cl/~ipedraza/

Schniederjans, M., Hoffman, J., Sirmans, G.: Using goal programming and the analytic hierarchy process in house selection. J. Real Estate Finan. Econ. **11**(2), 167–176 (1995)

Schoijet, M.: Capítulo 7. Las represas y sus efectos sobre la salud. En SENER, En Una introducción a la problemática de los impactos. Secretaría de Energía, México (1984)

Schwab, R.: Twenty Years of Policy Recommendations for Indigenous Education: Overview and Research Implications. Australian National University, Australia (1995)

SENER: Prospectiva del Sector Eléctrico 2013–2027. 2013. Secretaría de Energía, México (2013)

SENER: Estrategia Nacional de Energía 2014–2028. Secretaría de Energía, México (2014)

SENER: Programa de Desarrollo Eléctrico 2019–2033. Secretaría de Energía, México (2019)

SIE [software de computación]: SENER: Sistema de Información Energética (2010). Obtenido de http://sie.energia.gob.mx/

Suárez, R., Peirano, M.: Impactos socio ambientales de las mega represas. El caso Garabi: Fundación Moot. Recuperado el 24 de 03 de 2020 (2010). http://www.mbigua.org.ar/uploads/RepresaGarabi2010.pdf

Tajbakhsh, M., Memarian, H., Shahrokhi, Y.: Analyzing and modeling urban sprawl and land use changes in a developing city using a CA-Markovian approach. Glob. J. Environ. Sci. Manage. **2**(4), 397–410 (2016). https://doi.org/10.22034/GJESM.2016.02.04.009

Viera, A., Garrett, J.: Understanding interobserver agreement: the kappa statistic. Fam. Med. **37**, 360–363 (2005)

WEB-Hipre: Global decision support (Version 1.22) [software de computación]. Helsinki University of Technology: Systems Analysis Laboratory (2007)

Wichapa, N., Khokhajaikiat, P.: Solving multi-objective facility location problem using the fuzzy analytical hierarchy process and goal programming: a case study on infectious waste disposal centers. Oper. Res. Perspect. **4**, 39–48 (2017)

Yurdakul, M.: Selection of computer-integrated manufacturing technologies using a combined analytic hierarchy process and goal programming model. Robot. Comput. Integr. Manuf. **20**(4), 329–340 (2004). https://doi.org/10.1016/j.rcim.20

A State of the Art Review of Buckling Restrained Brace: History, Application, and Design

Hrishikesh Shedge[1], N. K. Patil[1], Anand Tapase[1](✉), Digvijay Kadam[1], Ajay Shelar[1], and Sudarshan Bobade[2]

[1] Rayat Shikshan Sanstha's, Karmaveer Bhaurao Patil College of Engineering, Satara, India
{nagendra.patil,digvijay.kadam,ajay.shelar}@kbpcoes.edu.in
[2] PCET's Pimpri Chinchwad College of Engineering and Research, Ravet, Pune, India

Abstract. A recent development in the field of seismic resistant structures is the buckling-restrained brace (BRB) which is a passive hysterics damper. Its uniform hysteric performance is achieved via a composite structure. Concrete is encased around a yielding geometry to control its buckling. A steel member is used as a core, concrete as casing material, and a stiff connector to incorporate this deceptive device in-between frames. The symmetric response of the core under tension and compression is achieved by restraining the core under critical buckling load. Due to its relatively simple design methodology and working BRBs are used in new as well as old structures as a retrofit. This paper reports the modeling and design procedure of BRBs along with their development since its first application. The design methodology and procedure for mathematical modeling of BRB are described in detail. The paper is a state of the art review based on the recent developments in BRBs and their application in structures. A focus on gap analysis underlining the loopholes and further requirements is addressed and are presented in the form of conclusions so that the future scope for the researchers can be underlined.

Keywords: Buckling · Restrained braces · Review

1 Introduction

The use of passive hysteretic dampers is long been done [1, 5]. A recent development in the field of seismic resistant structures is the buckling-restrained brace (BRB) which is a passive hysteric damper. Its uniform hysteric performance is achieved via a composite structure. Concrete is encased around a yielding geometry to control its buckling. A steel member is used as a core, concrete as casing material, and a stiff connector to incorporate this deceptive device in-between frames. A symmetric response of the core under tension and compression is achieved by restraining the core under critical buckling load. Due to its relatively simple design methodology and working BRBs are used in new as well as old structures as a retrofit. A stable hysteretic damper is proven to dissipate energy in the tension and compression region by its geometry. BRB is an amalgamation of the geometric and material syndicate in such a manner that it dissipates energy uniformly.

J. Neves et al. (Eds.): GeoChina 2021, SUCI, pp. 43–52, 2021.
https://doi.org/10.1007/978-3-030-80155-7_3

Frames incorporated with BRBs, known as BRBF, have proven to be highly efficient in damping the induced forces in the structure. BRBs since their invention in the 1990s have gone through a significant number of improvements and still, its refinement is under process [27–32]. Due to its wide horizon of application, BRB are begin used for steel structures as well as RCC multistory structures [2, 4] (Fig. 1 and Table 1).

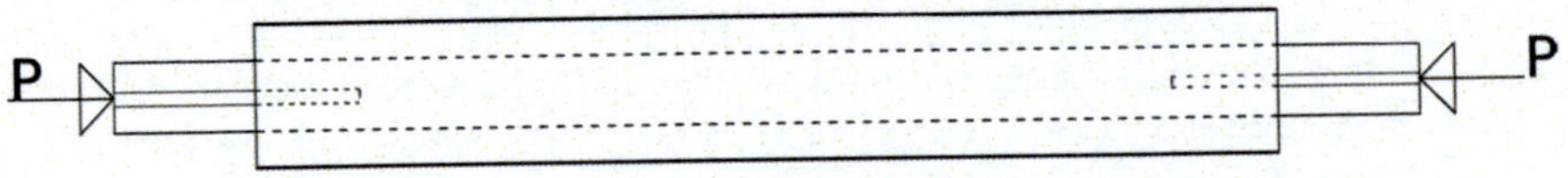

Fig. 1. Section of BRB

Table 1. List of places equipped with BRB in USA [5]

Location	Yield strength	Number of braces
Plant and Environmental Sciences Building—University of California, Davis, California	115–550 kips	132 braces
Marin County Civic Center Hall of Justice—County of Marin, California	Py = 400–600 kips	44 braces
Broad Center for the Biological Sciences—California Institute of Technology, California	Py = 285–660 kips	84 braces
Hildebrand Hall — University of California, Berkeley, California	Py = 200–400 kips	36 braces
Wallace F. Bennett Federal Building—Federal General Services Administration, Salt Lake City, Utah	Py = 205–1905 kips	344 braces
Building 5, HP Corvallis Campus—Hewlett-Packard, Corvallis, Oregon	Py = 110–130 kips	60 braces
Centralized Dining and Student Services Building— University of California, Berkeley, California	Py = 210–705 kips	95 braces
King County Courthouse—King County, Seattle, Washington	Py = 200–500 kips	50 braces
Genome and Biomedical Sciences Building—University of California, Davis, California	Py = 150–520 kips	97 braces
Physical Sciences Building—University of California at Santa Cruz, California	Py = 150–500 kips	74 braces

(continued)

Table 1. (*continued*)

Location	Yield strength	Number of braces
Second Research Building (Building 19B)—University of California, San Francisco, California	Py = 150–675 kips	132 braces
Kaiser Santa Clara Medical Center Hospital Building Phase I, Kaiser Permanente, Santa Clara, California	Py = 265–45 kips	120 braces

2 Literature Review

Larry et al. [2] studied the large scale buckling restrained brace frames to understand the ductility demand in realistic structures. Their observations led to the conclusion that BRBF exhibit poor performance at story drift between 0.02–0.025 rad. They also concluded that the large stiffness of beam-column brace geometry results in large flexural demand and in turn cause the catastrophic failure of the frame. They concluded that a properly designed BRB can sustain without significant stiffness degradation up to story drift of 0.05 rad and ductility demand of close to 25.

Zaid A et al. [23] conducted an experimental study on the application of the new type of BRB as a retrofit for seismically deficient R.C.C frames. Researchers developed a novel end connector that allowed for expansion and contraction while providing lateral restraint against buckling. The researchers also used three different types of materials as core and found that steel core provides the best performance compared to others. Cigdem et al. [3] modeled a nonlinear finite element analysis of steel core BRB and aluminum core BRB. Their numerical model from Ansys- Workbench gave results in cognizance to full-scale experiments carried on BRB frames by the author. The researchers were able to model the FE parameters to get the realistic behavior of R.C.C BRBF. Marshall et al. [15], in their research, highlighted the use of BRB for steel frames. The author also described the importance of BRB in terms of ease of design and application. Xie [22], has summarized the importance of BRB and its application. Xie has also given the required design parameters and differential equations governing the behavior of BRB. Wada et al. [1], did pioneering work in developing BRB and its methodology for commercial use. The author has done rigorous work in developing the correlation between core and encasing concrete. The author also concluded that the encasing steel tube should have an Euler buckling load 1.5 times higher than that of the core. Their study also found that the concrete used for encasing contributed a significant amount of flexural stiffness. The authors also observed the stable hysteric performance of BRB. Yazdi et al. [21], have replaced concrete with steel plate assembly as an encasing material. The author has used a steel plate assembly to restrict buckling encasing and has performed a parametric FE analysis in ABAQUS. This new generation BRB is called steel BRB and is lighter than the conventional BRB. The authors concluded that the critical load ratio of BRB and core should not be less than 1.5. The author has found out that the gap between core and ensuing causes the geometric instability in higher modes under cyclic load. Black

C. et al. [5], explained in detail the design and features of commercially used buckling restraining brace, known as, "Un-bonded Brace", developed by Nippon steel Japan. The authors explained the phenomena of flexural buckling, buckling of the core at higher modes, plastic torsional buckling of et al., and de-bonding of materials. The authors have also carried out experimental work by testing 5 large scale BRBs. The authors found out the reliability of the brace and its significances in improving the performance of the beam-column frame. Gaetano et al. [7], designed and installed an all-steel buckling restrained brace (BRB) in a damaged R.C.C two-story building. The BRB was designed by researchers in such a manner that bulging of core occurred before buckling of core. The designed BRB was able to withstand the story drift of 0.03 rad and showed the stable hysteretic performance. P.P Rossi [9], explained the use of Isotropic hardening for modeling of buckling restraint brace. Multiple models of BRBs lead the author to the conclusion that the model which considered both isotropic hardening and kinematic hardening is likely to produce inaccurate results. Clark et al. [4], explained the design and application of large-scale buckling restraining brace manufactured commercially by Nippon steel. The authors also evaluated the performance of BRBF as per guidelines given by FEMA. The authors have also discussed the structures incorporated with BRB within the USA. Makris N. [14], explained the plastic torsional buckling of the cruciform column. The author highlighted its importance in the design of BRB. The author also explained how the incremental theory of plasticity governs the onset of torsional buckling. Zub C. et al. [23] developed two types of buckling restrained braces, where authors tested conventional type BRB and a core stiffened BRB using numerical modeling. The authors' findings suggest that BRB having a rectangular cross-sectional area underperform as compared to BRB with a square cross-section. Pandikkadavath M. et al. [24] developed and tested a hybrid buckling restraining brace to enhance its performance under cyclic loading. Hoveidae N. and Rafezy B. [25], analyzed all-steel buckling restrained brace to study interface details and magnitude of friction between core and casing. Wang C. et al. [26], analyzed a modular BRB and checked its performance under low cycle fatigue.

3 Differential Equation Governing Stability of BRB Under Axial Compression [21]

DE is given by Euler and is found in the book by Temoshinko. For a bar under axial compressive force P, length l and modulus of rigidity EI

$$U = \frac{EI}{2}\int_0^l \left(\frac{d^2y}{dx^2}\right)dx \tag{I}$$

The equilibrium equation for uniform transverse reaction offered by the encasing mortar in the deformed configuration is given as

$$E_i I_i \frac{d^4y}{dx^4} + P\frac{d^2y}{dx^2} = -q \tag{II}$$

−q, indicates that the reaction offered by mortar is in opposite direction to that of deflection.

DE in terms of equal and opposite distributed load q, offered by encasing mortar is given as

$$E_o I_o \frac{d^4 y}{dx^4} = q \tag{III}$$

Substituting in the above equation

$$E_i I_i \frac{d^4 y}{dx^4} + P \frac{d^2 y}{dx^2} = -E_o I_o \frac{d^4 y}{dx^4} \tag{IV}$$

A homogeneous Euler buckling load equation is obtained after rearranging the terms

$$\frac{d^4 y}{dx^4} + \frac{P}{(EiIi + EoIo)} \frac{d^2 y}{dx^2} = 0 \tag{V}$$

Critical buckling load is then given by

$$Pcr = \frac{\pi^2}{(kL)^2}(EiIi + EoIo) \tag{VI}$$

Since the Modulus of rigidity of the inner steel core is small compared to that of the encasing mortar.

Critical buckling load is thus stated in form of modulus of rigidity of encasing motor alone

$$Pcr = \frac{\pi^2}{(kL)^2}(EoIo) \tag{VII}$$

A stable brace has Pcr > Py.

4 The Relation Between Elastic Modulus of Casing/Mortar and Core Buckling Stress [21]

It can be found in the book Temoshinko and Gere, where they have energy-based relation for a bar buckling on an elastic foundation. The same analogy is applied for the buckling core of BRB and its surrounding casing.

The quantity β is the modulus of encasing/mortar and P is the axial force causing buckling.

Then, the expression for deflection of the curve with both ends hinged can be derived in terms of the Fourier series

$$y = a1 \frac{\sin \pi x}{l} + a2 \frac{\sin 2\pi x}{l} + a3 \frac{\sin 3\pi x}{l} + \ldots \tag{1}$$

The Strain Energy of bending of the core is given by equation

$$\Delta Ui = \frac{EI}{2} \int_0^l \left(\frac{d^2 y}{dx^2}\right)^2 dx \tag{2}$$

Substituting the above equation in it

$$\Delta Ui = \frac{\pi^4 EI}{4l^3} \sum_{n=1}^{n=\infty} n^4 a_n^2 \tag{3}$$

The strain energy of the encasing/mortar can be found by integrating lateral reaction offered by element dx of the casing, which turns out to be $(\beta y^2/2)\ dx$. The total energy of deformation is then calculated by integrating the energy of element over length l.

$$Uo = \frac{\beta}{2} \int_0^l y^2 dx \tag{4}$$

Substituting Eq. (1) in (4) gives total strain energy of casing

$$Uo = \frac{\beta l}{4} \sum_{n=1}^{n=\infty} a_n^2 \tag{5}$$

Hence the total strain energy of the assembly is given as;

$$U_T = Ui + Uo \tag{6}$$

And work done by the compressive axial force on the core is equal to;

$$Wp = \frac{P\pi^2}{4l} \sum_{n=1}^{n=\infty} n^2 a_n^2 \tag{7}$$

In a stable hysteretic brace $U_T = Wp$;

$$\frac{\pi^4 EI}{4l^3} \sum_{n=1}^{n=\infty} n^4 a_n^2 + \frac{\beta l}{4} \sum_{n=1}^{n=\infty} a_n^2 = \frac{P\pi^2}{4l} \sum_{n=1}^{n=\infty} n^2 a_n^2 \tag{8}$$

For parametric equation in terms of axial load can be plotted as;

$$P = \frac{\pi^2 EI}{l^2} \frac{\sum_{n=1}^{n=\infty} n^4 a_n^2 + \frac{\beta l^4}{\pi^4 EI} \sum_{n=1}^{n=\infty} a_n^2}{n^2 a_n^2} \tag{9}$$

Critical buckling load P_{cr} is calculated from the above equation by minimizing the coefficient a to a_m that is, where the curvature of the bar is simple and the value of a_m is greater than zero.

$$P_{cr} = \frac{\pi^2 EI}{l^2} \left\{ m^2 + \frac{\beta l^4}{m^2 \pi^4 EI} \right\} \tag{10}$$

Where m is an integer, which gives the number of half-sine waves present in the core when the force reaches a critical value. EI is the modulus of rigidity of the core.

Now, β is calculated for value m, where m transitions from 1 to 2,

$$\beta = \frac{4\pi^4 EI}{l^4} \tag{11}$$

And a relation between P_{cr} and β can be derived by substituting β in Eq. (9)

$$P_{cr} = 2\sqrt{\beta EI} \tag{12}$$

For designing BRB and to keep in check the higher modes of buckling, $P_{cr} \geq \sigma_y A$.

From which a parametric relation between β and various cross-sections of BRB is developed.

$$\beta \geq \frac{\sigma_y^2 A^2}{4EI} \tag{13}$$

When the core is rectangular with bt cross-sectional area

$$\beta \geq 3\frac{\sigma_y^2 b}{Et} \tag{14}$$

When the core is cruciform in shape with $2bt$ cross-section area

$$\beta \geq 12\frac{\sigma_y^2 t}{Eb} \tag{15}$$

The above relation implies that the rectangular core needs a higher value of β to avoid local buckling. β calculated from the above method gives answers which differ in magnitudes from the experimental results. The reason being, concrete is sandwiched between core and outer steel casing. This encasing of concrete mortar causes concrete to offer more lateral restrain. This phenomenon on increase in lateral stiffness of encased material is evaluated using plane strains.

β, is constrained modulus or the stiffness of the encased mortar. Using constitutive equations for one-dimensional analysis, the fourth orders tress tension C_{ijkl} turns out to be equal to β

$$\beta = \frac{E_c(1-\mu_c)}{(1+\mu_c)(1-2\mu_c)} \tag{16}$$

Where E_c and μ_c are moduli of elasticity and Poisson's ratio of encased concrete respectively.

The Relation Between Core Steel and Encasing Tube [1]

For an unbonded brace, there is no transfer of axial force from the core to the encasing tube, the equilibrium of assembly can be achieved by keeping the buckling load of casing slightly higher yield load of the core.

In reality, this equation is influenced by various parameters likes end condition, initial deflection, material non-uniformity, etc. By considering the initial deflection due to self-weight Wada et al. [1], gave a relation between axial load P and buckling load Pe (Fig. 2).

$$\frac{P_E}{P_y} = 1 + \left(\frac{\pi^2 Ee}{2\sigma_{y(tube)} l}\right) \Big/ \left(\frac{l}{D}\right) \tag{17}$$

The experiments performed by Wada showed that the ratio $\frac{P_E}{P_y} = 1.03$ was sufficient to avoid that buckling of the core instead of the actual ratio of 1.5. This was because of encased concrete when contributed to additional flexural stiffness.

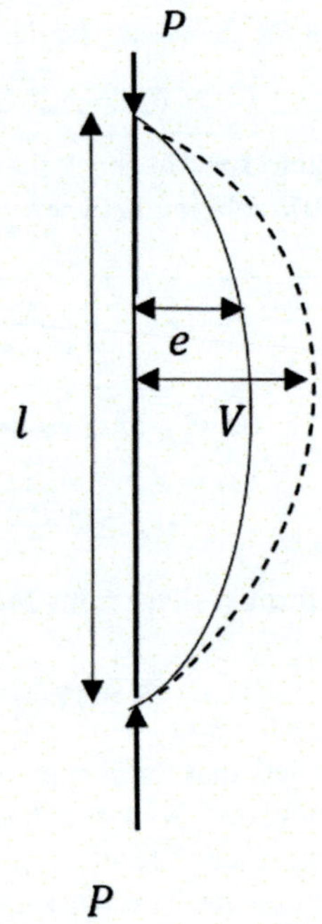

Fig. 2. Buckling of Slender Section Column

5 Conclusion

Use of Buckling Restrained Brace for performance enhancement of structure under cyclic loading has proven to be efficient and beneficial. The ease of design and application has made BRB a popular choice over conventional passive hysteretic dampers. The current research in BRB is focused on new-age materials, which will enhance its performance and overcome current limitations.

References:

1. Akira, W., Masayoshi, N.: Infancy to maturity of buckling restrained braces research. In: 13th World Conference on Earthquake Engineering, Vancouver, pp. 1–6, Paper No. 1732 (2004)
2. Andrews, B.M., Song, J., Fahnestock, L.A.: Assessment of buckling-restrained braced frame reliability using an experimental limit-state model and stochastic dynamic analysis. Earthq. Eng. Eng. Vibr. **8**(3), 373–385 (2009). https://doi.org/10.1007/s11803-009-9013-8
3. Avci-Karatas, C., Celik, O.C., Eruslu, S.O.: Modeling of buckling restrained braces (BRBs) using full-scale experimental data. KSCE J. Civ. Eng. **23**(10), 4431–4444 (2019). https://doi.org/10.1007/s12205-019-2430-y
4. Clark, P., Aiken, I., Ko, E., Kasai, K., Kimura, I.: Design procedure for building incorporating hysteretic damping devices. In: Proceedings of the 68th Annual Convention, Santa Barbara, California. Structural Engineers Association of California (1999)

5. Black, C., Markris, N., Aiken, I.: Component testing, stability analysis and characterization of buckling- restrained unbonded braces. PEER Report (2002)
6. Farouk, M.A., El-Kady, M.S.: Mathematical model of analyzing P–δ effect in braced columns. Innov. Infrastruct. Solutions **3**(1), 1–15 (2018). https://doi.org/10.1007/s41062-018-0147-2
7. Corte, G.D., D'Aniello, M., Landolfo, R.: Field testing of all-steel buckling-restrained braces applied to a damaged reinforced concrete building. https://doi.org/10.1061/(ASCE)ST.1943-541X.0001080.
8. Ghowsi, A.F., Sahoo, D.R.: Seismic performance of nine-story self-centering buckling-restrained braced frames. In: Prakash, R.V., Suresh Kumar, R., Nagesha, A., Sasikala, G., Bhaduri, A.K. (eds.) Structural Integrity Assessment. LNME, pp. 801–813. Springer, Singapore (2020). https://doi.org/10.1007/978-981-13-8767-8_68
9. Rossi, P.P.: Importance of isotropic hardening in the modeling of buckling restrained braces. J. Struct. Eng. **141**(4), 04014124 (2015). https://doi.org/10.1061/(ASCE)ST.1943-541X.0001031
10. Jia, M., Guo, L., Lu, D.: Performance testing and comparison of buckling-restrained braces with H and crisscross cross section unrestrained segments. Int. J. Steel Struct. **14**(4), 745–753 (2014). https://doi.org/10.1007/s13296-014-1206-y
11. Jiang, T., Dai, J., Yang, Y., Liu, Y., Bai, W.: Study of a new-type of steel buckling-restrained brace. Earthq. Eng. Eng. Vib. **19**(1), 239–256 (2020). https://doi.org/10.1007/s11803-020-0559-9
12. Fahnestock, L.A., Ricles, J.M., Sause, R.: Experimental evaluation of a large-scale buckling-restrained braced frame. J. Struct. Eng. **1205–1214**, 133–139 (2007). https://doi.org/10.1061/(ASCE)0733-9445
13. Markis, N.: Plastic torsional buckling of cruciform compression members. J. Eng. Mech. (ASCE) **129**, 689–696 (2003). https://doi.org/10.1061/(ASCE)0733-9399
14. Marshall, J.D.: Buckling-restrained braces and their implementation in structural design of steel buildings. In: Beer, M., Kougioumtzoglou, I., Patelli, E., Au, I.K. (eds.) Encyclopedia of Earthquake Engineering. Springer, Heidelberg (2021). https://doi.org/10.1007/978-3-642-36197-5_313-1
15. Narayan, K.K., Pathak, : Buckling analysis of braced frames under axial and lateral loadings: the effect of bracing location. In: Adhikari, S., Dutta, A., Choudhury, S. (eds.) Advances in Structural Technologies: Select Proceedings of CoAST 2019, pp. 317–334. Springer, Singapore (2021). https://doi.org/10.1007/978-981-15-5235-9_24
16. Pham, D.-H., Chou, C.-C.: Test of a full-scale two-story steel X-BRBF: strong-axis instability of buckling restrained brace associated with out-of-plane bending of gusset connections. In: Reddy, J.N., Wang, C.M., Luong, V.H., Le, A.T. (eds.) ICSCEA 2019. LNCE, vol. 80, pp. 375–380. Springer, Singapore (2020). https://doi.org/10.1007/978-981-15-5144-4_32
17. Qiu, C., Zhu, S.: Enhance seismic performance of self-centering concentrically braced frames by using hybrid systems. Bull. Earthq. Eng. **18**(8), 3995–4015 (2020). https://doi.org/10.1007/s10518-020-00851-x
18. Sangtarash, H., Banan, M.R., Banan, M.R.: Performance evaluation of a new NSP for estimation of seismic demands of nonlinear irregular brb frames. Iran. J. Sci. Technol. Trans. Civ. Eng. **44**(3), 803–812 (2019). https://doi.org/10.1007/s40996-019-00275-x
19. Shankar, H.J.P., Lamsal, S., Shrestha, P., Ganesh, B., Prabhakara, R.: Performance evaluation of concentric and eccentric buckling restrained braces on the dynamic behaviour of RC structures. In: Vinyas, M., Loja, A., Reddy, K.R. (eds.) Advances in Structures, Systems and Materials. LNMIE, pp. 249–257. Springer, Singapore (2020). https://doi.org/10.1007/978-981-15-3254-2_23
20. Timoshenko, G.: Theory of Elastic Stability, 2nd edn. Mcgraw-Hill International Book Company (1963)

21. Xie, Q.: State of the art of buckling-restrained braces in Asia. J. Constr. Steel Res. **61**, 727–748 (2005). https://doi.org/10.1016/j.jcsr.2004.11.00523]
22. Yazdi, H.M., Mosalman, M., Soltani, A.M.: Seismic study of buckling restrained brace system without concrete infill. Int. J. Steel Struct. **18**(1), 153–162 (2018). https://doi.org/10.1007/s13296-018-0312-7
23. Zub, C.-I., Stratan, A., Dogariu, A., Vulcu, C., Dubina, D.: Development of two types of buckling restrained braces using finite element modelling. In: Vacareanu, R., Ionescu, C. (eds.) Seismic Hazard and Risk Assessment. SNH, pp. 373–387. Springer, Cham (2018). https://doi.org/10.1007/978-3-319-74724-8_25
24. Pandikkadavath, M.S., Sahoo, D.R.: Development and subassemblage cyclic testing of hybrid buckling-restrained steel braces. Earthq. Eng. Eng. Vib. **19**(4), 967–983 (2020). https://doi.org/10.1007/s11803-020-0607-5
25. Hoveidae, N., Rafezy, B.: Local buckling behavior of core plate in all-steel buckling restrained braces. Int. J. Steel Struct. **15**(2), 249–260 (2015). https://doi.org/10.1007/s13296-015-6001-x
26. Cao, X., Gang, W., Feng, D.-C., Wang, Z., Cui, H.: Research on the seismic retrofitting performance of RC frames using SC-PBSPC BRBF substructures. Earthq. Eng. Struct. Dyn. **49**(8), 794–816 (2020). https://doi.org/10.1002/eqe.3265
27. Zhu, B.-L., Guo, Y.-L., Gao, J.-K., Pi, Y.-L.: Behavior and design of spatial triple-truss-confined BRBs with a longitudinal shuttle shape. Eng. Struct. **215**, 110605 (2020). https://doi.org/10.1016/j.engstruct.2020.110605
28. Wang, C.-L., Zhao, J., Gao, Y., Meng, S.: Experimental investigation of modular buckling-restrained energy dissipaters with detachable features. J. Constr. Steel Res. **172**, 106191 (2020). https://doi.org/10.1016/j.jcsr.2020.106191
29. Barbagallo, F., Bosco, M., Marino, E.M., Rossi, P.P.: Achieving a more effective concentric braced frame by the double-stage yield BRB. Eng. Struct. **186**, 484–497 (2019). https://doi.org/10.1016/j.engstruct
30. Freddi, F., Tubaldi, E., et. al.: Seismic performance of dual systems coupling moment-resisting and buckling-restrained braced frames (2020). https://doi.org/10.1002/eqe.3332
31. Qie, Y., Barbagallo, F., et al.: Full-scale hybrid test for realistic verification of a seismic upgrading technique of RC frames by BRBs. Earthq. Eng. Struct. Dyn. **49**(14) (2020). https://doi.org/10.1002/eqe.3312
32. Atasever, K., Inaaga, S., et al.: Experimental and numerical studies on buckling restrained braces with post-tensioned carbon fiber composite cables. Earthq. Eng. Struct. Dyn. **49**(15) (2020). https://doi.org/10.1002/eqe.3321

Characteristics of Lithely (Flexible) Arch Bridges and Case Studies from Satara

Digvijay Kadam(✉), N. K. Patil, M. Anand Tapase, Ajay Shelar, and Hrishikesh Shedge

Rayat Shikshan Sanstha's, Karmaveer Bhaurao Patil College of Engineering, Satara, India
{digvijay.kadam,nagendra.patil,ajay.shelar}@kbpcoes.edu.in

Abstract. Arch bridges are phenomenal structures constructed world over advantageously considering their durability, strength, pleasant aesthetics, and nearly negligible maintenance cost over many years. It is learned that the time-consuming process of constructing an arch bridge with the wedge-shaped precise cuts makes it costly in the initial construction. Considering this drawback, the Government of Maharashtra has proposed to construct the arch bridges with the modern lithely arch method comparatively cost-effective and fastest than the conventional method. Recently lithely arch method of the bridge is used in several places for the rehabilitation or construction of a new arch bridge in the Satara district of Maharashtra state in India. Precast wedge-shaped blocks are used in the lithely method reducing the construction time. In the present paper characteristics of lithely arch bridges along with several case studies from Satara district are discussed about their rapid installation and cost-effectiveness including details of the various approaches to construction.

Keyword: Lithely Arch Bridge

1 Introduction

Masonry arch bridges were successfully constructed without carrying any analytical technique due to their unavailability in past. The effectiveness of empirical design methods has stood strong since the beginning of the masonry arch bridges era. Even though the analytical and empirical methods have been applied since the initial period of the 18th century, in ordinary places the tedious analytical procedures fail and are found too difficult to analyze. Engineering judgment i.e. the use of empirical methods wherein the engineering plans were not available is allowed by the AASHTO Manual for the Condition Evaluation of Bridges (AASHTO 1994).

The method states that if the bridge withstands at certain loading without any failure, it was presumptively adequate to sustain that type of load. Compression is the dominant stress induced in the arch under uniform loading is the main advantage of an arch bridge [5]. So the low cost with high compression strength constructed using the materials like stone and concrete best suits to get a shape of an arch. Small and medium span bridges are constructed using such heavy materials were constructed using a full shoring system.

J. Neves et al. (Eds.): GeoChina 2021, SUCI, pp. 53–65, 2021.
https://doi.org/10.1007/978-3-030-80155-7_4

Advances in the use of high strength concrete, steel, and concrete-steel composites in recent years have significantly reduced the weight of the structure and have extended the limits of arch bridges to longer spans. The masonry arch bridges were the popular types of bridges until the invention of advanced materials which have brought a revolution in making a variation in the design of bridges [2]. But even today the arch bridges remain in use with the help of modern and advanced technology and materials.

Fast durable constructions along with the construction of longer spans are the added advantages of the advent of new technologies and materials [6]. Lithely Arch is a precast, modular, concrete arch bridge system. The bridges constructed using Lithely Arch technology is purely compressive. Lithely Arch technology can be used for replacing, widening, and strengthening old arch bridges. Masonry arches are strong, durable, aesthetically pleasing, and largely maintenance-free, yet since 1900 there has been a dramatic decline in their use [1]. However, designers, contractors, and clients now have access to a new method of constructing arches incorporating precast concrete voussoirs interconnected via polymeric reinforcement and a concrete screed. No centering is necessary, as the Flexi Arch when it is lifted, transforms under the forces of gravity into the desired arch shape.

Masonry arch bridges are one of the oldest railway infrastructures that serve the rail industry even in modern times. Despite their age, these bridges are sturdy, elegant, and serve society and industry with minimal maintenance. Lack of education and research in structural masonry, the loss of skilled masons to construct these arches, and the emergence of modern materials after post-war industrialization have caused concern for the safety and serviceability of these bridges to asset owners. Lithely means the flexible term first introduced in the UK. In Asia, it is first introduced in India, Maharashtra. Lithely (Flexible) arch method is the fastest, economical method of arch bridge construction. A bridge constructed using this method is purely compressive. This method utilizes precast units for the construction of arch bridges. This method can be utilized for the rehabilitation of old arch bridges. In the present study, the emphasis is given to studying the construction of a lithely arch bridge through a case study of Kadhane.

2 Literature Review

Michael et.al showcased the good blending of the classic and organic look of a masonry arch bridge that allows for the structure with its natural surroundings, furthermore, the paper presents the rehabilitation and strengthening of two different types of masonry arch bridges. Solla, et.al stated the advantageous use of Ground-penetrating radar (GPR) to analyze the bridge its state of conservation from a historical, archaeological, and structural point of view. Hongrui et.al. presented the advantages of the rehabilitation scheme. Ozden et.al stated three-dimensional finite element analysis of a case study of a bridge constructed in the earthquake prompt area. Jan et.al presents a method for the restoration and rehabilitation of Liben Bridge, a historic concrete vaulted arch bridge structure from the early 20th century.

3 Overview of Study Area

The Lithely arch bridge site selected as a case study is from village Kadhane, Post Dhebewadi; Taluka Patan is located in the Satara District of Maharashtra State, India. Figure 1 gives the details of the location map of the selected case study. The bridge site is 17° 14′53.8" N 74° 02′05.0"E in Maharashtra state which is 610 m above sea level. The lithely arch bridge is constructed in place of the old bridge site as shown in Fig. 2 and Fig. 3.

4 Materials and Methods Used for the Construction of the Lithely Arch Bridge

The bridge is manufactured and taken to the site in flat-pack form using polymeric reinforcement to carry the self-weight of the arch unit during lifting but once in place, it behaves as a conventional masonry arch. The method of construction and material used for its manufacturing is shown in Fig. 2. For the manufacture of each arch unit the tapered voussoirs are precast individually and then they are laid closely with the top periphery touching, in a horizontal line with a layer of polymeric reinforcement placed on top. In situ screed, typically 40–50 mm thick, is placed on top and allowed to harden so that the voussoirs are interconnected.

When lifted at the designated anchorage points, gravity forces cause the wedge-shaped gaps to close, concrete hinges form in the screed, and the integrity of the unit is provided by the tension in the polymeric reinforcement and by the shear resistance of the screed. The arch-shaped units are then lifted and placed on precast footings at the bridge site and all the self-weight is then transferred from tension in the polymeric reinforcement to compression in the voussoirs, wherein it acts in the same way as a conventional masonry arch. The voussoirs can be accurately, quickly, and consistently produced with the desired taper is relatively simple shuttering, and high-quality concrete is utilized to enhance the durability of the arch while in service and greatly reduce the variability associated with natural stonework. Site experience has shown that a typical unit can be accurately located every after 15 to 20 min and as a consequence, most bridges can be installed in a day leading to rapid installation making the method affordable and convenient over a conventionally constructed arch and making the system competitive with beam and slab alternatives.

It has also been found that bearings are not required to allow longitudinal expansion reducing the maintenance requirements related to the bearing which needs to be replaced during the lifetime of the bridge. This allied to the lack of corrodible reinforcement means that maintenance will be minimal and their design lives should be well more than alternatives.

Different grades of concrete and other materials are used in the parts of the bridge for the construction of lithely including for the construction of blocks M_{50} grade concrete, for precast blocks PCC M_{50}, for a pier, abutment, foundation PCC M_{30}, leveling course -PCC M_{10}, annular filling PCC M_{15}, screed concrete PCC M_{50}, plum concrete for back feel PCC M_{15}, load distributers PCC M_{25}, approach slap RCC M_{25}, wearing course

RCC M_{30}, curb RCC M_{20}, railing G.I. steel collapsible, drainage spouts 100 mmØ G.I. pipe, spandrel cladding C.C brick, retaining wall RCC M_{30}.

Figure 4(a) shows the Mould used for the casting of precast blocks. The sizes of moulds are depending upon the size of the block. The sizes of the block are depending upon the height and span of the bridge. Figure 4(b) shows metal rails used during screed concreting of precast blocks for the leveling purpose of the block strip. The length of the metallic rail is 3000 mm & the spacing between the two rails is 500 mm of each. Figure 4(c) shows the wire rope used for laying the precast blocks in a line. Depending upon the number of blocks 16 mm or 20 mm wire ropes are used. Figure 4(d) shows ferrule through which wire ropes are passes with an 8 mm *varnothing* bar of 100 mm length. It is a machine pressed against the wire rope along with the bar at the spacing of 610 mm. The bar placed in ferrule is bent up at 900 for anchorage. Figure 4(e) shows the Silicone Resipro chemical which is acts as a sealant in between the two blocks. Figure 4(f) shows a separation plate which is provided to separate the screed concreting of each block. It is also having provision to pass the wire ropes through it & also provided with handles for lifting it. Figure 4(g) shows a metallic Z- shaped plate which is used to reduce the size of screed concrete by 50 mm from all edge of the blocks. Installation of the lithely arch bridge on the site of Kadhane, post-Dhebewadi, Satara is various stages is shown in Figs. 5, 6, 7, 8 and 9 sequentially. The Lithely arch bridge site selected as a case study is from village Koregaon - Kedareshwar, Post Koregaon; Taluka Koregaon is located in the Satara District of Maharashtra State, India. Figure 4 gives the details of the location map of the selected case study. The bridge site is 17°41′47.2"N 74°09′30.8"E in Maharashtra state which is 610 m above sea level. The lithely arch bridge is constructed near to the old bridge site as shown in Figs 10, 11, 12 and 13. In both case studies which are from the village are for which we do have readily available traffic loading conditions. For the village area we usually consider class A-A loading in the first case which is the major road connecting village Kadhane, Post Dhebewadi; Taluka Patan to the main road of Patan. In the second case study, we usually consider class B-B type loading which is an internal village road in Koregaon.

5 Observations and Discussions

The Lithely arch method is very useful for the winding of the old arch bridge, rehabilitation of the bridge. Rehabilitation of arch bridge carried out by lithely arch method is initially felt costlier than other types of rehabilitation but after performing life cycle cost study it proved economical. Rehabilitation of the arch bridge carried out by replacing the bridge with a lithely arch method is more economical than rehabilitation carried out by replacing the bridge with the RCC beam bridge method. The Lithely arch method is an easy, fastest, and economical method than other methods of construction of the arch bridge. No use of bearing and other attachments so this method ensures zero maintenance. There is no chance of corrosion due to no use of reinforcement. This bridge has a life of above 100 years. During the construction of a lithely arch bridge, no centering is necessary.

Fig. 1. Location map of the study area at Kadhane, post-Dhebewadi, Satara.

Fig. 2. Location of study area bridge site

Fig. 3. Old bridge location where Lithely Arch Bridge constructed

Fig. 4. Materials and methods used for the construction of the lithely arch bridge

Fig. 5. Strips of concrete blocks used for laying on the rail for leveling

Fig. 6. Lifting of the block strip using a crane

Fig. 7. Mould used for the preparation of block

Fig. 8. After lifting of block strips

Fig. 9. Series of Arches are placed

Fig. 10. Location map of the study area at Kadhane, post-Dhebewadi, Satara

Fig. 11. Location of study area bridge site

Fig. 12. Old bridge at site

Fig. 13. Construction of new bridge near old bridge

References

1. Long, A., Kirkpatrick, J., Gupta, A., Nanukuttan, S., Polin, D.M.: Rapid construction of arch bridges using the innovative FlexiArch. Proc. Inst. Civil Eng. Bridge Eng. **3**, 143–153 (2013)
2. Xie, W., Zhao, T., Tang, J., Zhang, Y.: Arch first and beam later: arch-rib integral installation construction technology for large-span tied-arch bridge. J. Constr. Eng. Manage. **143**(8) ((2017) https://doi.org/10.1061/(ASCE) CO.1943–7862.0001356. © 2017 American Society of Civil Engineers. Pages 04017059(1) – 04017059-(8)
3. Sarmiento-Comesías, M., Ruiz-Teran, A.M., Aparicio, A.C.: Structural Behavior of Inferior-Deck Spatial Arch Bridges with Imposed Curvature. J. Bridge Eng., ASCE **17**, 682–690 (2012)
4. Dhanasekar, M., Prasad, P., Dorji, J., Zahra, T.: Serviceability assessment of masonry arch bridges using digital image correlation. J. Bridge Eng. **24**(2), 04018120 (2019). https://doi.org/10.1061/(ASCE)BE.1943-5592.0001341. © 2018 American Society of Civil Engineers. Pages 04018120(1)–04018120(16)
5. Liu, Z., Li, F., Kim Roddis, W.M.: Analytic model of long-span self-shored arch bridge. J. Bridge Eng. ASCE, 14–21 (2002). https://doi.org/10.1061/(ASCE)1084-0702(2002)7:1(14)
6. Joshi, S.V., Kanade, G.N.: Rehabilitation of arch bridges with lithely arch method. Int. J. Adv. Res. Dev. **4**(1), 48–49 (2019)
7. IRC: 6–2016 "Standard Specifications and Code of Practice for Road Bridges" Indian Roads Congress, New Delhi
8. Long, A., Nanukuttan, S.: Arch Bridges – Unlocking Their Potential. Journal of Engineering and Computational Mechanics, ICE Proceedings. (2016). https://doi.org/10.1680/jencm.16.00018
9. Long, A., Gupta, A., Mc Polin, D.C., Cook, J.: Adapting the flexiarch for widening a complex arch bridge. In: Proceedings of the ICE - Bridge Engineering (2018)
10. Ura, A., Oruc, S., Dogangu, A., Tuluk, I.: Turkish Historical Arch Bridges And Their Deterioration's And failures. Int. J. Eng. Failure Anal., Elsevier Ltd. 43–53 (2008)

11. Lahti, M.A., P.E., Leo A. Fernandez, P.E. : "Rehabilitation of Masonry Arch Bridges", ASCE Structures Congress 2010. Orlando, Florida, United States. (2010). https://doi.org/10.1061/41130(369)290
12. Solla, M., Lorenzo, H., Riveiro, B., Rial, F.I.: Non-destructive methodologies in the assessment of the masonry arch bridge of Traba, Spain. Eng. Failure Anal. **18**(3), 828–835 (2011). https://doi.org/10.1016/j.engfailanal.2010.12.009
13. Liu, H., Zou, J., Jiang, H., Lin, G.: Structure analysis of continuous composite arch bridge for rehabilitation. Second International Conference on Transportation Engineering (2012). https://doi.org/10.1061/41039(345)107
14. Ozden Caglayan, B., Ozakgul, K., Tezer, O.: Assessment of a concrete arch bridge using static and dynamic load tests. Struct. Eng. Mech. **41**(1), 83–94 (2012). https://doi.org/10.12989/sem.2012.41.1.083
15. Teigen, J.G.: Rehabilitation Design of a Historic Concrete Arch Bridge in Prague from the Early 20th Century. J. Performance Constructed Facilities **34**(4) (2020). https://doi.org/10.1061/(ASCE)CF.1943-5509.0001451

The Design Parameters and Quality Requirements of Jet Grout Columns in the Stabilization of a Sloping Bermed Excavation

Arthur K. O. So(✉)

AKOS Geotechnical Consulting Limited, Wan Chai, Hong Kong

Abstract. Jet grouting is used worldwide but is uncommon in Hong Kong. Few local examples can be found which are used for soil improvement and ground water control. This project is the first time to use jet grout columns (JGCs) to stabilize a sloping bermed excavation for the construction of a basement structure by pseudo top-down method in a reclaimed land. Original design was to install several rows of 2m diameter JGGs at 4 m c/c spacing, serving as dowels, through the marine deposits and alluvium with 2 m into reclamation fills above and 2 m into completely decomposed granite (CDG) below. Diameter and strength are two primary design parameters. Numerous studies and field applications are established for the prediction, but their reliability is strongly influenced by the uncertainties in the mixing uniformity of the grouted materials. Random strength is always produced irrespective of the soil type. The knowledge of jet grouting in reclamation fills and CDG is also inadequate. Trial JGCs were therefore installed, aiming to determine the operation parameters of the jetting and to improve the mixing uniformity of the grouted materials. Proof cores revealed that debris such as artificial materials, soil lumps and cobbles from the reclamation fills above were collected at the bottom of the JGCs and inadequate treatment in the CDG due to the strength and bonding nature of the weathered soil were frequently observed. The design was then revised to 0.5 m nominal embedment into CDG to reflect the quality of JGCs achieved. Working columns were installed and 28 days proof cores were taken to determine the total core recovery, unconfined compression strength, Young's modulus and failure strain. These values were also found varying largely due to the large variability in the strength of the treated soils and the non-homogeneous nature of the JGCs formed by jet grouting. The results are statistically analyzed and compared with other researchers. In view of the large variation of material properties within the JGC, it is important for the designer to understand how the design system works and how the JGCs are mobilized in order to determine the JGC quality requirements and to optimize the design.

Keywords: Sloping bermed excavation · Jet grout column · Soil improvement · Dowel · Design intent · Design parameters · Operation Parameters · Quality requirements

J. Neves et al. (Eds.): GeoChina 2021, SUCI, pp. 66–84, 2021.
https://doi.org/10.1007/978-3-030-80155-7_5

1 Introduction

Jet grouting is an in-situ injection technique employed with specialized equipment that include grout pump, grout mixer, drill rig, drill rods and injection monitor with horizontal radial nozzles delivering high velocity fluids to erode, mix, and stabilize in-situ soils using an engineered grout slurry (GEO Institute 2009). Columnar element of grouted soil, also named as soilcrete, is produced. Common applications are underpinning, foundation, tunnel protection, excavation support, slope stabilization and ground water sealing. It was developed in Japan in 1970s, first applied in United States in the early 1980s and gradually became popular worldwide (Burke and Meffe 1991, van der Stoel 2001, Shibazaki, 2003, Brill et al. 2003, Ho 2005). The jet grouting system can be distinguished into three basic types in operation; namely the single fluid (grout only), double fluid (grout and air) and triple fluid (grout, air and water). A simplified comparison between them is given in Table 1.

Table 1. A simplified comparison between the three systems (Siepi 2014) * lower * * intermediate *** higher

Characteristics	Single fluid	Double fluid	Triple fluid
Column diameter	*	***	***
Increase in strength	***	*	**
Reduction in permeability	***	**	***
Original soil replacement (%)	*	**	***
Quantity of spoil rising to surface	*	**	***
Cost of equipment	*	**	***
Risk of heave at surface	***	**	*
Productivity	*	***	***

With continuous development and advancement in the jet grouting techniques over recent years, new variations have emerged (Brill et al. 2003, Burke 2012, Wang et al. 2013, Atangana et al. 2018), such as the Super Jet (Yoshida et al. 1996), Cross Jet (Yoshida and Saito 2009), Rodin Jet Pile (Nakanishi et al. 1997, Wang et al. 2013), Twin-Jet (Shen et al. 2013a) and Metro Jet (Nakashima and Nakanishi 1995), and the elliptical jet grouting (Leoni and Pianezze 2017) and horizontal jet grouting (Atangana et al. 2017). The type of system to be used is determined by the prevailing soil conditions, geometrical form and required quality of jet grouting. As different types of soils may exhibit different erodibility characteristics (Burke 2012), successful performance of the grouting is highly dependent on the operation parameters in relation to the achievable column diameters, mixing uniformity, strength and deformation properties of the grouted materials.

The achieved diameter is difficult to determine due to the inherent variability of soil conditions, and the complex interaction between jet grout and soil such that the cutting energies in different soil types can be substantially different. According to Ho (2005), the state of practice and research at his time suggested that the methods of predicting the diameter are still highly empirical and rely on a data base of field trial results to assist in the selection of the operation parameters. Some examples are given in Table 2. The Schlosser (1997) method is widely used because of its simplicity.

Table 2. Predictions of the jet grout column (JGC) diameter

Methods	Equations	References
Empirical approach	$D \alpha E_s$ $E_s = pQ/v$	Schlosser (1997)
	$D \alpha E_s$ $E_s = E_{sg} = p_g Q_g/v$ $= E_{sg} + E_{sw} = p_g Q_g/v + p_w Q_w/v$	Paggliacci et al. (1994)
	$D = Kp^{\alpha} Q^{\beta} N^{\gamma}/v^{\delta}$	Shibazaki (2003)
	$D = n_1 (pQ/v)^{n}{}_2$	Mihalis et al. (2004)
Theoretical approach	$D = 1.25 d_o\ ((p - p_o)/q_u)^{1/2} + D_r$	Ho (2005)
	$D = 2\int V_c t\ dt$	Modoni et al. (2006)

Note:-
D = JGC diameter
E_s = specific energy supplied by the jet to form a unit height of JGC
E_{sg} = specific energy contributed by grout (in single fluid system) or grout and air (in double fluids system)
E_{sw} = specific energy contributed by water (in triple fluids system)
p = grout pressure supplied by the jet to form a unit height of JGC
P_g = grout pressure contributed by grout (in single fluid system) or grout and air (in double fluids system)
pw = grout pressure contributed by water (in triple fluids system)
Q = flow rate of grout supplied by the jet to form a unit height of JGC
Qg = flow rate of grout contributed by grout (in single fluid system) or grout and air (in double fluids system)
Qw = flow rate of grout contributed by water (in triple fluids system)
v = lifting speed of jetting rod
N = number of revolutions
K = a coefficient of soil type, 31.5 for sandy soils
α, β, γ, δ = empirical coefficients = 1.003, 1.186, 0.135 and 0.198 respectively
n1, and n2 = empirical coefficients
Vc = penetration rate of the fluid jet in soil
t = duration of action of the jet on soil
dt = infinitesimal time
do = nozzle diameter
po = presiding pressure at nozzle outlet
qu = ultimate bearing resistance of soil
Dr = diameter of monitor

In this decade, various approaches on predictive models have developed as shown in Table 3.

Table 3. Various approaches on predictive models in this decade

Categories	Authors	Comments
Conceptual	Arson and Juge (2012)	Applicable to single-, double- and triple-fluid systems
Theoretical	Wang et al. (2012)	
	Shen et al. (2013b)	
Semi-theoretical	Flora et al. (2013)	
Artificial intelligence	Ochmanski et al. (2015)	
	Tinoco et al. (2016)	
Measurement	Bearce et al. (2015)	Direct current electrical resistivity push probe
	Cheng et al. (2017)	Acoustic monitoring
	Lin et al. (2019)	In-hole electrical resistivity tomography

According to Kauschinger et al. (1992), van der Stoel (2001) and Gladkov et. al. (2011), strength is controlled by the cement content and soil types. Modoni and Bzowka (2012) found that the Young's modulus E_c can be correlated to the unconfined compression strength (UCS or σ_c) of the cemented body as $E_c = \beta\sigma_c$ where β is the correlation coefficient ranging from 300 to 400. Akan et al. (2015) predicted the UCS and determined the effect of grouting parameters using multiple regression analysis of the grouting pressure, rotation-lifting speed, water-cement ratio and water flow rate. Wanik et al. (2017) proposed a relationship between the UCS of the jet-grouted materials, soil types and quantity of cement as shown in Fig. 1. Toraldo et al. (2017) introduced a methodology to quantify the uncertainties, characterizing the different factors of variability in the jet-grouted material and simulating their effects on the representative structural elements based samples cored from the columns and sonic tomography scans performed on large blocks. Repeated systematic analysis of the variable parameters leads to general formulae expressing the statistical distribution of the UCS, which can be combined to calculate the correcting factors to give the characteristic strength of the grouted materials.

Jet grouting is uncommon in Hong Kong. Few local examples can be found in recent years which are used for soil improvement and ground water control as shown in Table 4.

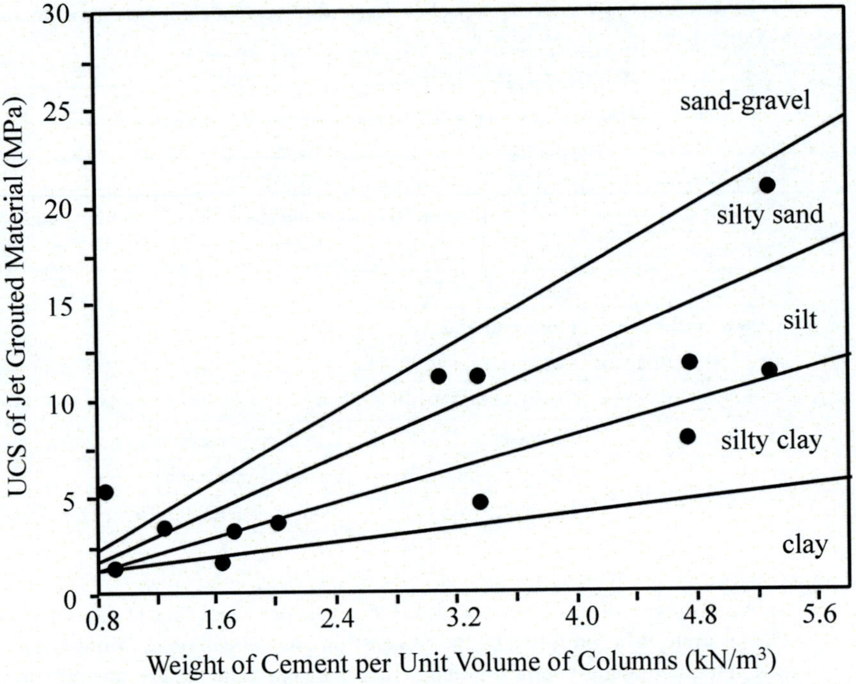

Fig. 1. Relationship between UCS, soil types and cement content (Wanik et al. 2017)

The presented project is the first time to use jet grout columns (JGCs) serving as dowels to stabilize a sloping bermed excavation for the construction of a basement structure in a reclaimed land. Diameter and strength of the JGCs are two primary concerns in the design. Numerous studies and field applications have established for their prediction. However, their reliability is still strongly influenced by the uncertainties in the mixing uniformity of the grouted materials. Furthermore, the knowledge of jet grouting in reclamation fills and completely decomposed granite (CDG) is inadequate. Site trial was therefore carried out to determine the operation parameters of the jetting and to improve the mixing uniformity of the grouted materials. Proof cores revealed that debris such as artificial materials, soil lumps and cobbles from the reclamation fills were collected at the bottom of the JGCs and inadequate treatment in CDG were frequently observed. The design was therefore revised to reflect the quality of JGCs achieved. Working columns based on the revised design were then installed and 28 days proof cores were taken to determine the total core recovery (TCR), UCS, Young's modulus (E) and failure strain. These values are also found varying largely due to the large variability in the strength of the treated soils and the non-homogeneous nature of the JGCs formed by jet grouting. The results are statistically analyzed and compared with other researchers. The design parameters and the quality of JGCs are discussed and some suggestions in relation to the design intents are made.

Table 4. Use of jet grouting in Hong Kong

Projects	Applications
Hong Kong-Zhuhai-Macau Bridge (Hong Kong boundary crossing facilities)	To treat the weak soils below the joint section of an artificial island and the immerse tube tunnel to reduce the settlement of the tunnel
Hong Kong-Zhuhai-Macau bridge (section between Scenic Hill and Hong Kong boundary crossing facilities)	To serve as a watertight contact between the pile toes and bedrock
Hong Kong-Zhuhai-Macau bridge (passenger clearance building)	As a horizontal plug to provide a watertight bottom slab below the sea water pumphouse
Tai Lam Tunnel	To improve the soils for the installation of a pipe roof above the tunnel, along with rock bolts and shotcreating to stabilize the tunnel bore in extensive sections of weathered granite
Phase 3 central reclamation	As a support and cut-off curtain to close the gap of diaphragm walls or socketed H-piles in the excavation and lateral support system due to obstructions of utilities
Lai Chi Kok drainage tunnel	Soil improvement to facilitate the tunnel break out at the tunnel outfall where the soil is weak and very permeable
Shatin to Central Link – To Kwa Wan	Soil improvement to mitigate the settlement of existing building and structures due to diaphragm construction
Shatin to Central Link – Wanchai Station	As a support and cut-off curtain to close the gap of diaphragm walls or socketed H-piles in the excavation and lateral support system due to obstructions of utilities
West Kowloon terminus station North (This Project)	To stabilize the marine deposits and serve as an excavation and lateral support for the station in a reclaimed land

2 The Project Site

Ground Conditions

The project site is located on a reclaimed land with existing ground level at about + 4 mPD to +5.5 mPD and groundwater level fluctuating from about +0.8 mPD to + 1.8 mPD. As shown in Fig. 2, the geological stratigraphy comprises reclamation fills overlying the marine deposits (MDs), alluvium, residual soils and granitic bedrock. The MDs are very soft to firm silty clay with inter-bedded deposits of marine sand. The alluvium is firm or dense but locally soft or very soft. The interface between MDs and underlying alluvium is irregular. However, the site has a complex history of reclamation and use. As a result, the reclamation has given rise to disturbance of the MDs in the form of mud waves so that some very soft spots are found. In some locations, the MDs have been treated by surcharging, geotextiles and wick drains. The undrained strengths and thicknesses of the marine deposits are therefore very variable.

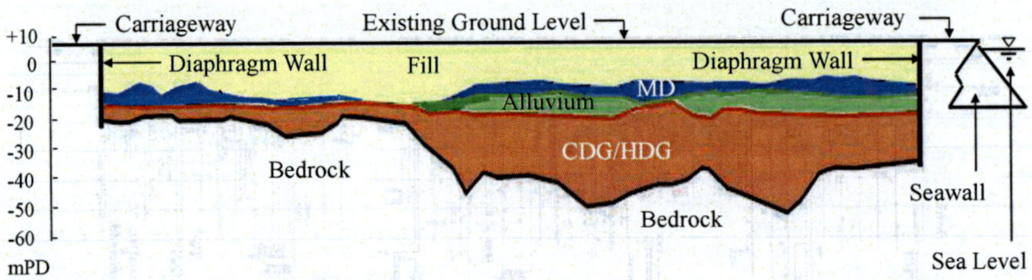

Fig. 2. Geological stratigraphy in N-S direction across the site (vertical exaggeration = 2)

Design

The basement structure is about 550 m long, 250 m wide and 30 m deep. It involved the installation of perimeter diaphragm wall which is to serve as a temporary support for bulk excavation and as a permanent basement wall when it is in service. This diaphragm wall also acts together with bored piles and socket H-piles as the piled foundation for the basement structure and the future building development. The perimeter diaphragm wall was designed and installed to "toe-in" to Cat 1(c) or better rock by 300 mm, with depth varying from about 20 m to 55 m depending on the inferred rockhead level. At the areas of high rockhead, the bases of the wall panels were provided with shear pins to guard against the kick-out instability and fissure grouting was carried out to form a hydraulic cut-off.

Construction

As shown in Fig. 3, a pseudo top-down construction or a central-island technique as named by others at the central part, and a full top-down construction at the north and south ends of the station were adopted. After installation of the diaphragm wall, bulk excavation took place in the form of a sloping bermed excavation, with 1:2 cut slopes at the north and south, and with 1:2 and 1:4 cut slopes supporting the diaphragm wall in the east and west. Bottom-up construction was then carried out at the central core, and by top-down construction at the periphery. As the stability of the cut slopes is inadequate in certain areas where the MDs are thick and/or weak, JGCs were installed to improve the stability of the cut slopes to the required factor of safety (FOS) of 1.2.

Design Parameters for the JGCs

Figure 4 is a schematic section of the sloping bermed excavation with several rows of 2 m diameter JGCs passing through the MD and alluvium layers, with 2 m into the fill lying above and 2m into CDG lying below. In the PLAXIS analysis, JGCs are modelled as non-porous elastic perfectly plastic columns with 2 m nominal embedment into CDG and maximum shear strength governed by the Mohr-Coulomb failure criterion.

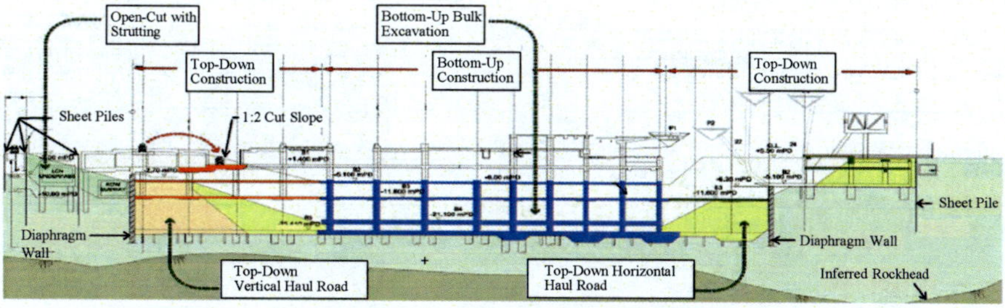

Fig. 3. Cross section in W-E direction across the basement

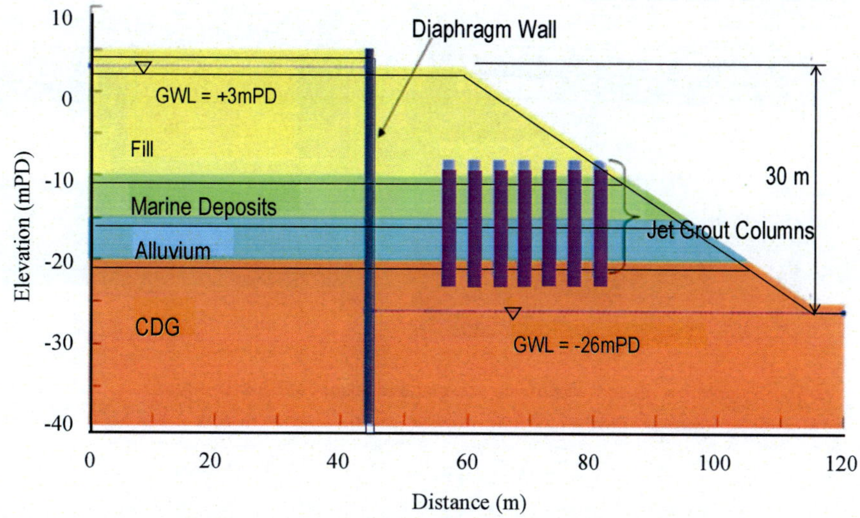

Fig. 4. Schematic section of a Bermed excavation stabilized by jet grout columns

To determine the operation parameters for jet grouting, the strength and stiffness of the JCGs are first specified for the design. The cement content and hence the w/c ratio for the grout mix and grout volume for a soil type are worked out. Based on The Schlosser (1997) method to predict the column diameter, the operation parameters for the grouting are determined. Trial columns are then installed and their qualities are verified by proof cores. If the design intents cannot be met, the operation parameters are adjusted. Figure 5 shows the proof cores of the first and second trials. Based on the trial JGCs test results, the operation parameters at the project site were determined as summarized in Table 5. The UCS and E of the JGCs adopted for the design were taken as 2 MPa and 300 MPa respectively. As the jet grouted material is quasi-brittle, 0.5% shear strain was adopted as the limiting criterion for the deformation. The proof coring results of the working JGCs are studied and compared with other researchers in the following section.

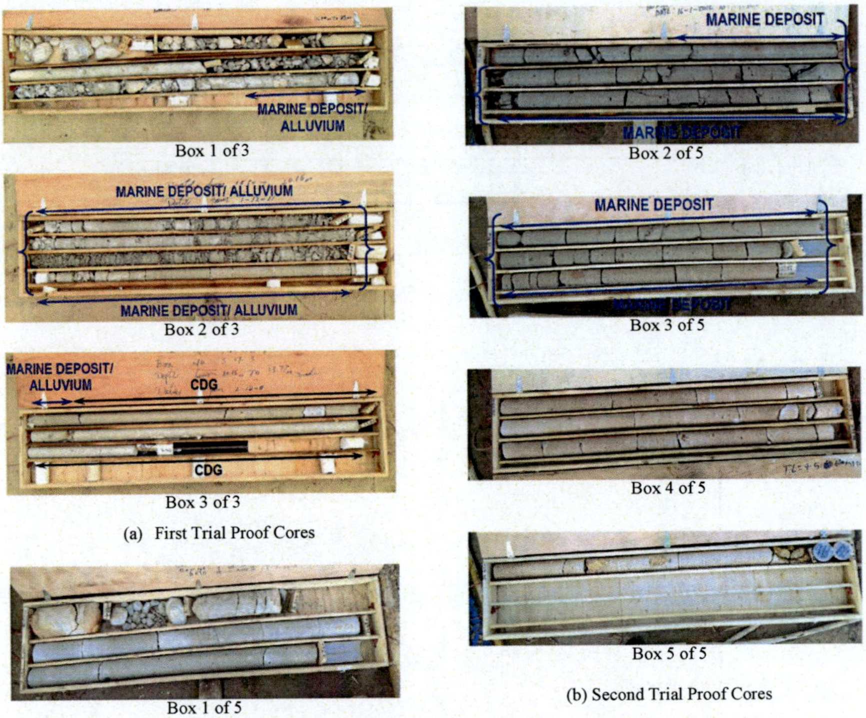

Box 1 of 3

Box 2 of 3

Box 3 of 3

(a) First Trial Proof Cores

Box 1 of 5

Box 2 of 5

Box 3 of 5

Box 4 of 5

Box 5 of 5

(b) Second Trial Proof Cores

Fig. 5. Representative proof cores of the trails

Table 5. Operation parameters in the trials

Operation parameters	First trial	Second trial
Replacement ratio	83%	88%
Water cement ratio	1:1	1:1
Nozzle size	2 × 35 mm	2 × 4.0 mm
Grout pressure P	400 bar	400 bar
Grout flow rate Q	260 L/min	340 L/min
Lifting rate V	9 cm/min	12.3 cm/min
Rate per minute RPM	5 rev/min	33 rev/min
Step	6 cm	4 cm
Grout input volume	2600 L/min	2750 L/min

3 Study of the Proof Coring Results of the Working JGC

Proof Coring

Not less than a JGC in 100 m^2 on plan was selected for proof coring. Cores were obtained at the centre, at 0.4 times the diameter and at 0.8 times the diameter of the JGC using double tube Mazier core-barrel to obtain undisturbed samples. Triple tube core barrel was used if hard materials are likely encountered. Samples were taken at about 1.5 m below the top, at the middle and at 1.5 m above the bottom of each core for the testing of the UCS and stiffness. Inclusions such as artificial materials, soil lumps and cobbles from the upper layers and inadequate treatment in the CDG due to the strength and bonding nature of the weathered soil were frequently observed of the early proof cores as shown in Fig. 6.

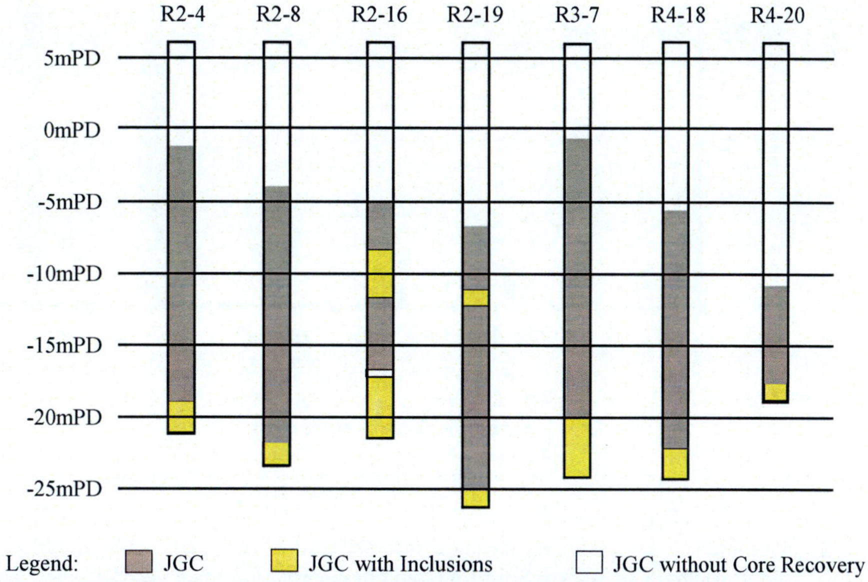

Fig. 6. Representative core logs

Total Core Recovery (TCR)

The general quality of the JGC is commonly assessed by the extent of grout treatment observed in the cores. Figure 7 is the statistical distribution of the TCR of 74 cores, of which the TCR taken at the edge of the JGC is generally less than that at the centre. This is consistent with the findings of Saurer et al. (2011) on the distribution of global properties of the improved soils. The average and standard deviation of TCR are determined equal to 82% and 25% respectively. According to the criteria given by GEO Institute (2009), a minimum TCR of 85% is required, implying that only 52 cores (74%) are satisfactory. TCR less than 85% may be caused by the poor workmanship in core drilling or may indicate the presence of untreated soft marine clay which could be flushed away during core drilling. On the other hand, TCR more than 85% may also be considered as unsatisfactory if it contains a large amount of marine clay. As the TCR of a core may give misleading interpretative result about the overall integrity of the JGC, TCR less than 85% locally shall not be adopted as a rigid and sole criterion for non-acceptance. Other factors such as the drilling rate of penetration and any observations during the core drilling process shall also be considered. In case of doubt, additional core can be taken and the JGC can be considered acceptable if the average TCR at all levels is greater than 85%. Nevertheless, the required quality of JGC should be reviewed in relation to its design intents and the probable failure mechanisms.

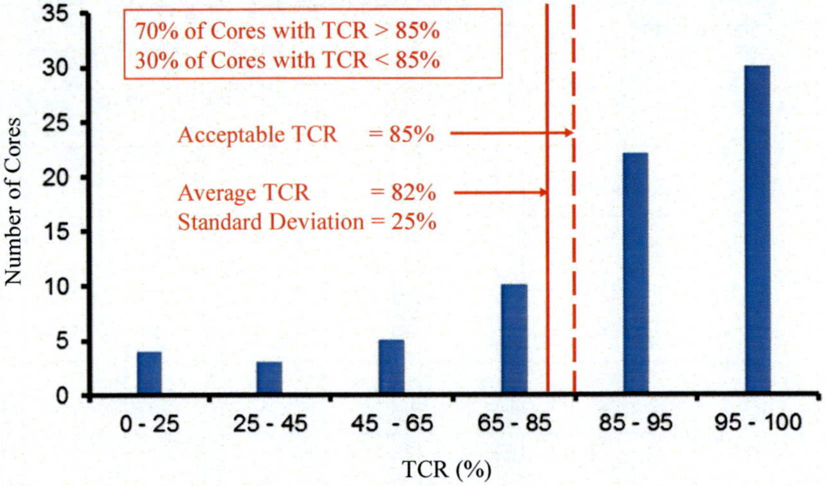

Fig. 7. TCR of proof cores

Unconfined Compression Test

Unconfined Compression tests were conducted following the British Standard BS1377-7:1990 "Methods of Test for Soils for Civil Engineering Purposes – Part 7: Shear Strength Tests (Total Stress)" (BSI 1999). Load frame method is used. Nominal specimen dimensions are 80 mm diameter and 160 mm long. Compression is applied to the specimen by motorized machine at a strain rate of 0.07 mm/min. Readings of the force-measuring device and the axial deformation gauges at regular intervals of compression are recorded simultaneously. At least 12 sets of readings are obtained. The test is continued until the maximum value of the axial stress has been passed. The force applied to the specimen for each set of readings is calculated by multiplying the change in readings of the force measuring device by a load calibration factor, while the axial strain δ of the specimen for each set of readings is calculated from the equation $\delta = \Delta L/L$ where ΔL is the change in length of the specimen and L is the initial length of the specimen. The compressive stresses are then calculated and plotted against the corresponding values of strains, and the stress-strain curve is defined by drawing a curve through the points. Failure is the point on the stress-strain curve at which the maximum compressive stress sustained by the specimen, and this point is used to determine the UCS and failure strain of the specimen. A typical stress-strain curve can be distinguished into three portions. The first portion is a gentle slope which is the result before the loading cell is in full contact with the testing specimen. The second portion is generally linear showing an elastic behaviour of the specimen and the E is determined from its slope. The third portion shows tension-softening after reaching the peak load. The test results of the JGC are discussed in the following. Typical UCSs of other soil types are given in Table 6, of which the UCS of treated clay is 3–7 MPa.

Table 6. Typical average unconfined compression strength of jet grouted soil (Kauschinger et. al. 1992)

Soil type	UCS (MPa)	
	Lower limit	Upper limit
Peat	1	6
Clay	3	7
Silt	5	15
Sand	10	40
Gravel	10	40

Unconfined Compression Strength (UCS)

Figure 8 shows the statistical distribution of the UCS of 364 core samples. The UCS varies from 0.79 MPa to 7.92 MPa with an average and standard deviation equal to 2.92 MPa and 1.17 MPa respectively; implying only 212 samples (84%) meet the design requirement of 2 MPa. The large scattering of the test results is likely due to the large variability in the strength of MDs and the non-homogeneous nature of the JGC formed by jet grouting. Furthermore, the representativeness of the UCS values to the overall quality of jet grouting may be arguable because UCS is always carried out on samples obtained from the most intact part of the cores, where the fractured material cannot be tested.

Young's Modulus (E)

Figure 9 shows the statistical distribution of the E determined from the stress-strain curve of the 364 UCS core samples. The E varies from 51 MPa to 1500 MPa, with an average and standard deviation equal to 430 MPa and 198 MPa respectively. This implies only 282 samples (77%) meet the design requirement of 300 MPa. E can also be obtained from the in-situ pressuremeter test, which in principle should eliminate the bias factor as in the UCS test because the pressuremeter test was carried out within the corehole at any depth. Indeed, pressuremeter test had been carried out near the centre of each trial JGC at 2 m interval for the whole depth of the column to determine the E. However, the test results are found largely scattered, with E ranging from 0 MPa to 1748 MPa, likely due to the limitation of the test, of which its accuracy highly depends on the experience and workmanship of the operators, and is therefore not used to analyze the E of the working JGC.

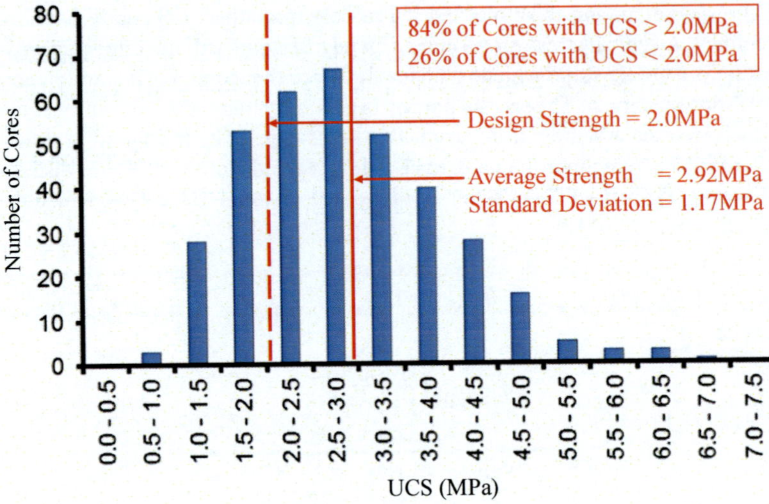

Fig. 8. UCS of core samples

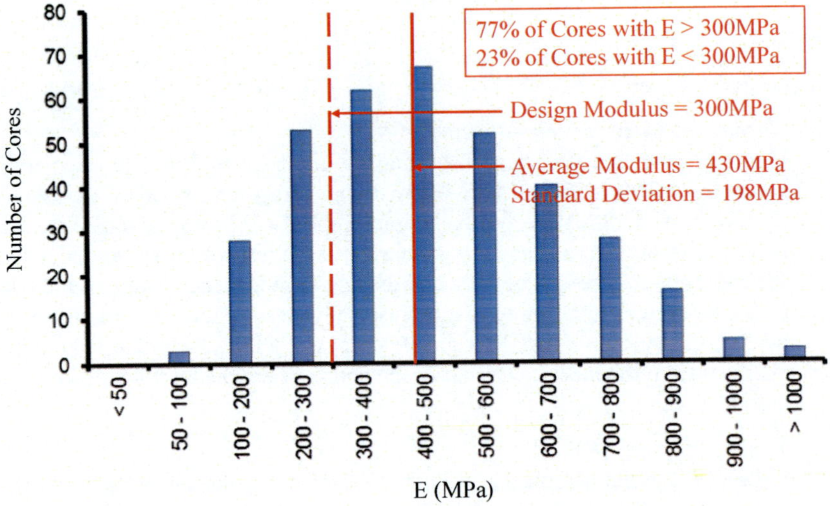

Fig. 9. Young's modulus of core samples

Failure Strain

Figure 10 is a typical stress-strain curve of a core sample under unconfined compression load. The UCS and the failure strain are determined as the compressive stress and axial strain at which the maximum stress sustained by the specimen, and the E is the slope in the elastic portion. The axial load is found experiencing a sudden drop at the peak load indicating a relatively brittle behaviour, then a diminishing residual strength with increased axial strain.

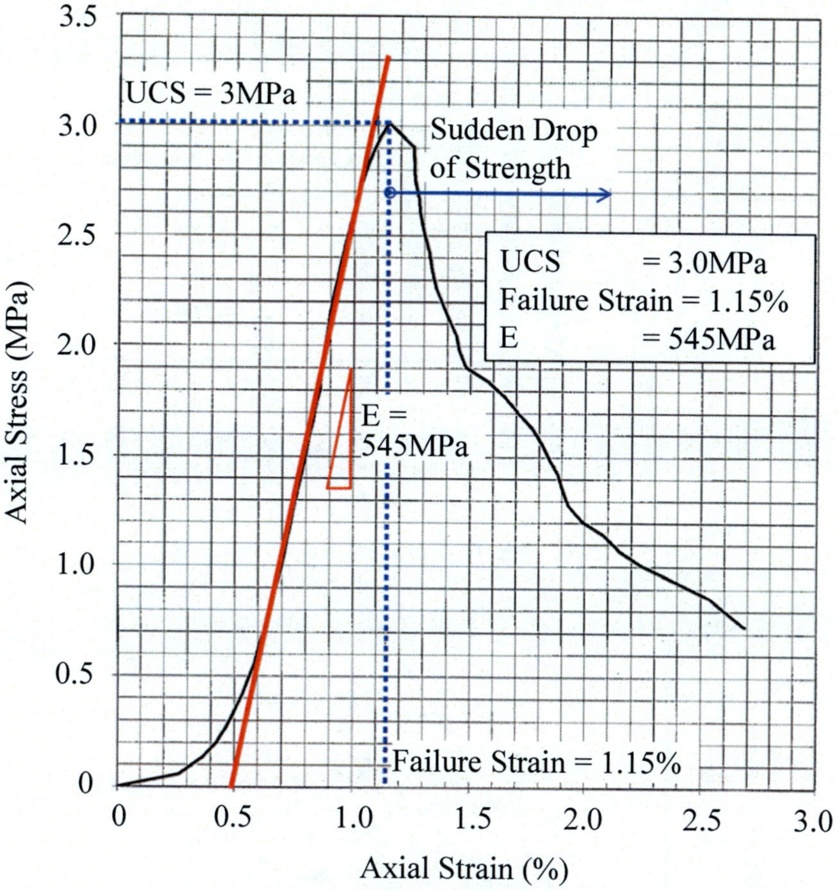

Fig. 10. Typical stress-strain curve

Figure 11 shows the statistical distribution of the failure strain determined from the stress-strain curve of the 364 UCS tests. The peak stress varies from 0.79MPa to 7.92MPa while the axial strain varies from 0.62% to 3.87% with an average and standard deviation equal to 1.46% and 0.52% respectively. This shows that most samples failed at a strain far beyond the design limit of 0.5%.

Correlation of E and UCS

Figure 12 shows the correlation of E and UCS of the 364 core samples. E is determined equal to 100–300UCS which is same as the JSG (1994) recommendation and an average E equal to 144UCS which is close to the design assumption of E = 150UCS. The large scattering of the test results is due to the large variability in the strength of MDs, the non-homogeneous nature of the JGC formed by jet grouting and the "eye-ball" judgment in the determination of the E value from the slope of the stress-strain curve.

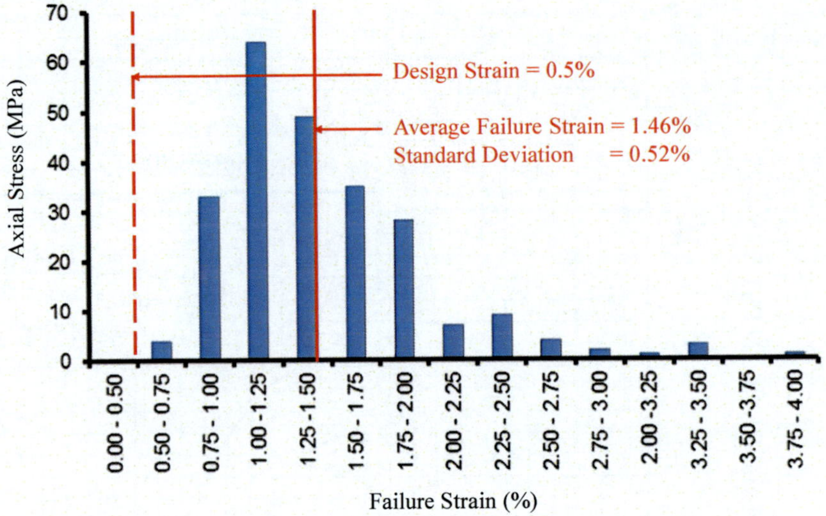

Fig. 11. Failure strain of core samples

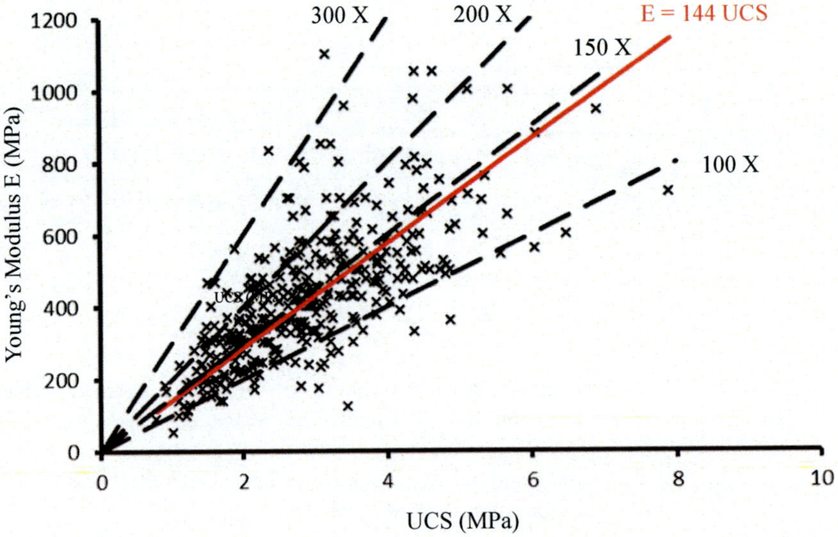

Fig. 12. Correlation of E and UCS

Summary of the Test Results and Comparison with Other Researchers

Table 7 is a summary of the test results and the test results of other researchers. It shows that the UCS, E, failure strains and correlation between E and UCS of the treated MDs are close to the test results of fine-grained soils obtained by other researchers.

Table 7. Summary of test results and comparison with other researchers

Projects		UCS (MPa)	E (MPa)	Failure strain (%)	Correlation between E & UCS
This project	Design	2	300	0.5	E = 150 UCS
	Test results	2.92±1.17	430±198	1.46±0.52	E = 144 UCS
JSG (1994)					E = 100 – 300 UCS
Kauschinger et al. (1992)		3–7			
van der Stoel (2001)	Peat Clay Eemclay	2.3±0.8 0.3±0.4 5.6±1.1	374±41 933±230 1532±427		E = 500 $(UCS)^{2/3}$
Jaritngam (2003)	Sand Silty soils Clay	10–30 10–30 1–10 2–6			E = 500 UCS E = 300 UCS E = 250 UCS E = 100 UCS
Shibazaki (2003)	Sandy cohesive	12.7 2.8			
Wen (2005)	Race course Road clarke Quay	1–10 1–7		0.5–1.5	E = 200 UCS
Axtell and Stark (2008)		3.24–10.83	345–1738	0.46–0.94	

4 Discussions and Conclusions

TCR of a proof core of not less than 85% is commonly specified as the acceptance criterion for the general quality of the JGC. Cautions in the core drilling are therefore required due to the quasi-brittle nature of the grouted materials. Furthermore, debris such as the artificial materials, soil lumps and cobbles collected from the upper fill layer and inadequate treatment in the CDG of the proof cores due to the strength and bonding nature of the weathered soil are frequently observed. Engineering judgement in the examination of the untreated materials in the core sample are required to ensure correct interpretative results. In the presented project, the design is revised from 2.0 m embedment into CDG to 0.5 m nominal embedment into CDG to reflect the quality of the JGCs achieved. Alternatively, the use of smaller diameter JGC or better control of the operation parameters to deliver higher energy than in the weaker soils can be used to improve the grouting quality in the CDG if the design requires a deeper penetration for better dowel capacity to limit the movements.

Histograms of UCS and E show large variance in values which are likely caused by the large variability in the strength of marine deposits and the non-homogeneous nature of the JGC formed by jet grouting. Such large variance in properties and materials is normally safeguarded by the use of lower bound values in the design which appears to give very conservative predictions. As JGCs are highly variable and quasi-brittle, limiting shear strain is normally adopted to avoid catastrophic failure. Likewise, the measured failure strains are much larger than the adopted values which also appear to imply that larger limiting value can be adopted. However, it shall be borne in mind that cores samples are always obtained from the intact part of the cores because fractured materials cannot be tested and therefore could have bias. The representativeness of the UCS, E and limiting shear strain to the overall quality of the jet grouting may post some uncertainty to the design.

In view of the large variation of material properties within the JGC, it is therefore important to understand how the design system works, i.e. mechanisms and failure mechanisms, and how the mass properties of the JGC are mobilized, e.g. in compression, shear, bending or their combinations. These are necessary information for the designer to determine the quality requirements and material parameters, and to optimize the design in relation to the embedment length if required, size, spacing and spatial arrangement e.g. in single, widely or closely spaced, buttress or block that can be chosen.

References

Akan, R., Keskin, S.N., Uzundurukan, S.: Multiple regression model for the prediction of unconfined compressive strength of jet grout columns. Procedia Earth Planet. Sci. **15**, 299–303 (2015)

Arson, C., Juge, B.: A hollow sphere model to dimension soilcrete columns. In: GeoCongress 2012. ASCE, Oakland (2012)

Atangana, N.P.G., Chen, J., Modoni, G., Arulrajah, A.: A review of jet grouting practice and development. Arab. J. Geosci. **11**, 459 (2018)

Atangana, N.P.G., Shen, J.S., Modoni, G., Arulrajah, A.: Recent advances in horizontal jet grouting (HJG): an overview. Arab. J. Sci. Eng. **43**(4), 1543–1560 (2017). https://doi.org/10.1007/s13369-017-2752-3

Axtell, P.J., Stark, T.D.: Increase in shear modulus by soil mix and jet grout methods. DFI J. **2**(1), 11–21 (2008)

Bearce, R.G., Mooney, M.A., Kessoun, P.: Estimation of jet grout column geometry using a DC electrical resistivity push probe. In: International Symposium on Non-Destructive Testing in Civil Engineering, Berlin, Germany, 15–17 September 2015 (2015)

Brill, G.T., Burke G.K., Ringen, A.R.: A ten-year perspective of jet grouting: advancements in applications and technology. In: Geotechnical Special Publication No. 120, Grouting and Ground Treatment, Proceedings of the 3rd International Conference, New Orleans, Louisiana, pp. 218–235 (2003)

BSI. BS1377-7:1990, Methods of Test for Soils for Civil Engineering Purposes – Part 7: Shear Strength Tests (Total Stress). British Standard Institute (1999)

Burke, G.K.: The state of practice of jet grouting. Grouting and Deep Mixing 2012, Geotechnical Special Publication No. 124, ASCE, Reston, pp. 875–886 (2012)

Burke, G.K., Meffe, D.A.: Fixing foundations. Civ. Eng. ASCE **61**(3), 63–65 (1991)

Cheng, S.H., Liao, H.J., Yamazaki, J., Wong, R.K.N.: Evaluation of jet grout column diameters by acoustic monitoring. Can. Geotech. J. **54**(2), 1781–1789 (2017)

Flora, A., Modoni, G., Lirer, S., Croce, P., Croce: The diameter of single, double and triple fluid jet grouting columns: prediction method and field trial results. Geotechnique **63**(11), 934–945 (2013)

GEO Institute. Jet grouting guideline. Geo-Instutute of ASCE Grouting Committee – Jet Grouting Task Force (2009)

Gladkov, I.L., Malinin, A.G., Zhemchugov, A.A.: Strength and deformation characteristics of soil-concrete as a function of jet-grouting parameters. In: Proceedings of the 21st European Young Geotechnical Engineer's' Conference Rotterdam, Netherlands, pp. 75–77 (2011)

Ho, C.E.: Turbulent fluid jet excavation in cohesive soil with particular application to jet grouting. PhD Thesis, Massachusetts Institute of Technology (2005)

Jaritngam, S.: Design concept of soil improvement for road construction on soft clay. In: Proceedings of the Eastern Asia Society for Transportation Studies (2003)

JSG. Jet Grout Design Manual, 6th Ed. Japanese Jet Grouting Association (1994)

Kauschinger, J.L., Perry, E.B., Hankour, R.: Jet Grouting: State of the Practice. Grouting, Soil Improvements and Geosynthetics – Geotechnical, vol. 30 no. 1, pp. 169–181. ASCE (1992) (Special Publications)

Leoni, F.M., Pianezze, G.: Elliptical jet grouting: an innovative, viable and effective solution: the example of bottom plugs for the SELA projects in New Orleans, LA. In: Grouting 2017, Jet Grouting, Diaphragm Walls and Deep Mixing, ASCE, Honolulu, pp. 11–20 (2017)

Lin, C.P., Lin, C.H., Ngui, Y.J., Wu, P.L.: Jet grouting column diameter measurement using in-hole electrical resistivity tomography. In: Proceedings of the 5th International Conference on Engineering Geophysics, Al Ain, UAE, 21–24 October, 2019 (2019)

Mihalis, I.K., Tsiambaos, G., Anagnostopoulos, A.: Jet grouting applications in soft rocks: the Athens metro case. In: Proceedings of Institution of Civil Engineer Geotechnical Engineering, 157 Issue GE4, pp. 219–228 (2004)

Modoni, G., Bzowka, J.: Analysis of foundation reinforced with jet grouting. J. Geotech. Geoenvironmental Eng. **138**(12), 1442–1454 (2012)

Modoni, G., Croce, P., Mongiovi, L.: Theoretical modelling of jet grouting. Geotechnique **56**(5), 335–347 (2006)

Nakanishi, W., Nakanishi, Y., Zhu, Q.L.: high pressure jet grouting method-RJP rodin jet pile and field practice in Beijing. China Saf. Sci. J. **7**(4), 35–42 (1997)

Nakashima, S., Nakanishi, W.: All-around type reinforcing and consolidating in the ground and apparatus thereof". United States, Patent Number: 5401121 (1995)

Ochmanski, M., Modoni, G., Bzowka, J.: Prediction of the diameter of jet grouting columns with artificial neural networks. Soils Found. **55**(2), 425–436 (2015)

Pagliacci, F., Trevisani, S., Chong, L.: Recent development in jet grouting techniques – Singapore Pulau Seraya power station, a case history. In: Proceedings of the 3rd International Conference on Deep Foundation Practice, Singapore, CI-Premier, pp. 219–225 (1994)

Saurer, E., Marcher, T., Lesnik, M.: Grid space optimization of jet grouting columns. In: Proceedings of the 15th European Conference on Soil Mechanics and Geotechnical Engineering, pp. 1055–1060 (2011)

Schlosser, F.: Soil improvement and reinforcement. In: Proceedings of the 14th International Conference on Soil Mechanics and Foundation Engineering, Hamburg, ISSMGE, pp. 2445–2466 (1997)

Shen, S.L., Wang, Z.F., Horpibulsuk, S., Kim, Y.H.: Jet-grouting with a newly developed technology: twin-jet method. Eng. Geol. **152**(1), 87–95 (2013)

Shen, S.L., Wang, Z.F., Yang, J., Ho, C.E.: Generalized approach for prediction of jet grout column diameter. J. Geotech. Geoenvironmental Eng. ASCE **2013**, 2060–2069 (2013)

Shibazaki, M.: State of practice of jet grouting. In: Geotechnical Special Publication No. 120, Grouting and Ground Treatment, Proceedings of the 3rd International Conference, New Orleans, Louisiana, pp. 198–217 (2003)

Siepi, M.: Choice of ground improvement techniques and design of ground improvement works. Seminar on Ground Improvement, Hong Kong, Centre for Research and Professional Development (2014)

Tinoco, J., Correia, A.G., Cortez, P.: Jet grout column diameter prediction based on a data-driven approach. Eur. J. Environ. Civ. Eng. **22**(3), 1–21 (2016)

Toraldo, C., Modoni, G., Ochmanski, M., Croce, P.: The characteristic strength of jet grouted material. Geotechnique **68**, 262–279 (2017)

van der Stoel, A.E.C.: Grouting for Pile Foundation Improvement. PhD Thesis, Delft University of Technology, the Netherlands (2001)

Wang, Z.F., Shen, S.L., Kim, Y.H.: Jet grouting practice: an overview. Geotech. Eng. J. SEAGS AGSSEA **44**(4), 88–96 (2013)

Wang, Z.F., Shen, S.L., Yang, J.: Estimation of the diameter of jet-grouting column based on turbulent kinematic flow theory. In: Grouting and Deep Mixing 2012, Geotechnical Special Publication No. 228, ASCE, Reston, pp. 2044–2051 (2012)

Wanik, L. Mascolo, M.C., Bzowka, J., Modoni, G., Shen, S.L.: Experimental evidence on the strength of soil treated with single and double fluid jet grouting. In: Proceedings of the Grouting: Grouting, Deep Mixing and Diaphragm Walls, Honolulu, 9–12 July (2017)

Wen, D.: Use of jet grouting in deep excavations. In: Ground Improvement Case Histories, pp. 357–370. Elsevier (2005)

Yoshida, H., Jimbo, S., Uesawa, S.: Development and practical applications of large diameter soil improvement method. In: Proceedings of 21st International Conference on Ground Improvement Geosystems: Grouting and Deep Mix (1996)

Yoshida, H., Saito, K.: Mechanism of cross jet and its application. In: Leung et al. (ed.) Ground Improvement Technology and Case Histories. Published by Research Publishing Service, Singapore, Geotechnical Society of Singapore (2009)

Performance of the Jet Grouted Sloping Berm as a Support to the Diaphragm Wall in an Excavation

Arthur K. O. So(✉)

AKOS Geotechnical Consulting Limited, Wan Chai, Hong Kong

Abstract. The use of jet grouting to stabilize a sloping berm in an excavation is uncommon. In a pseudo top-down construction project, after the installation of the diaphragm walls, bulk excavation took place at the central part with 1:2 sloping berms formed at the north and south ends, and with 1:2 and 1:4 sloping berms supporting the diaphragm walls at the east and west ends. Bottom-up construction was then carried out at the central core and by top-down construction at the periphery. As the site has a complex history of reclamation and land use, in areas where the marine clays are thick and/or weak, JGCs had to install to improve the stability of the sloping berms. Several rows of 2 m diameter JGGs at 4 m c/c spacing, serving as dowels, were installed through the marine deposits and alluvium with 2 m into reclamation fills above and 0.5 m nominal depth into completely decomposed granite below. Conventional design principles and parameters following the local codes of practices were adopted. The interaction of the diaphragm wall, slope and JGCs was simplified for a two-dimensional PLAXIS model analysis and the probable failure mechanisms were examined. The deflection of the diaphragm wall panels and JGCs, the ground water table in front of and behind the diaphragm walls were monitored as the excavation progressed. In this paper, a representative design section is presented and the performance of the lateral support system is compared with the design prediction. The actual movement of east wall is found larger than the conventional analysis of the movement around an excavation using values of elastic modulus equal to two times the elastic modulus of soil (E), i.e. 2E. This is consistent with the model findings that the mode of straining within the final form of the slope is predominantly by shearing rather than by direct straining. The deformation is therefore controlled largely by the value of shear modulus (G) within the slope, and the use of 2E would underestimate the predicted movement. Indifferently, the actual movement of west wall is close to the prediction using 2E because the sloping berm is wide and the mode of straining of the soil within the slope is predominantly by direct straining instead of by shearing.

Keywords: Jet grouted sloping berm · Jet grout columns · Piles on slope · Ground treatment · Failure mechanisms · Elastic modulus · Direct straining · Shearing · Rotation of principal stresses

1 Introduction

Jet grouting was developed in Japan in 1970s, first applied in United States in the early 1980s and gradually became popular worldwide (Burke and Meffe 1991, van der

J. Neves et al. (Eds.): GeoChina 2021, SUCI, pp. 85–101, 2021.
https://doi.org/10.1007/978-3-030-80155-7_6

Stoel 2001, Shibazaki, 2003, Brill et al. 2003, Ho 2005). Common applications are underpinning, foundation, tunnel protection, excavation support, slope stabilization and ground water sealing. With continuous development and advancement in the jet grouting techniques, new variations have emerged (Burke et al. 2012, Wang et al. 2013, Croce et al. 2014, Shen et al. 2014, Atagana et al. 2018). Main areas of researches in this decade are given in Table 1.

Table 1. Main areas of research in this decade

Research areas	Researchers
Jet grout column diameter prediction	Ribeiro and Cardoso (2017)
Jet grout column diameter measurement	Bearce et al. (2015) Cheng et al. (2017) Lin et al. (2019)
Uncertainties related to the strength variability	Modoni and Bzonka (2012) Akan et al. (2015) Wanik et al. (2017) Toraldo et al. (2017)
Variations in the jetting system	Burke (2012) Wang et al. (2013) Atangana et al. (2017)
Field applications	Ellis et al. (2010) Filz (2012) Zohrer (2017) Atangana et al. (2018) Bayesteh and Sabermahani (2020)
Construction effects	Yoshida (2012) Wang et al. (2014) Wu et al. (2016) Belleto et al. (2018)

The presented project is to use of jet grouting to stabilize a sloping berm in an excavation which is uncommon, and may be the first time in application. The site is located on a reclaimed land. Its geological stratigraphy comprises reclamation fills overlying the marine deposits (MDs), alluvium, residual soils and granitic bedrock. The MDs are very soft to firm silty clay with inter-bedded deposits of marine sand. As the site has a complex history of reclamation and use, it has given rise to disturbance of the MDs in the form of mud waves so that some very soft spots are found. In some locations, the MDs have been treated by surcharging, geotextiles and wick drains. The undrained strengths and thicknesses of the MDs are very variable. Design of the basement structure involved the installation of perimeter diaphragm wall which is to serve as a temporary support for bulk excavation and as a permanent basement wall when it is in service. The wall was designed and installed to "toe-in" to Cat 1(c) or better rock by 300 mm, with depth varying from about 20 m to 55 m depending on the inferred rockhead level. At the areas of high rockhead, shear pins were provided to the bases of the wall panels to guard against the kick-out instability and fissure grouting of rock below the bases of the panels to form a hydraulic cut-off.

Figure 1 shows a deep excavation for the basement construction. A pseudo top-down construction or a central-island technique as named by others at the central part, and a full top-down construction at the north and south ends of the station were adopted. After installation of the diaphragm wall, bulk excavation took place in the form of a sloping bermed excavation, with 1:2 cut slopes at the north and south, and with 1:2 and 1:4 cut slopes supporting the diaphragm wall in the east and west. Bottom-up construction was then carried out at the central core, and by top-down construction at the periphery. As the stability of the cut slopes is inadequate in certain areas where the MDs are thick and/or weak, JGCs were installed to improve the stability of the cut slopes to the required factor of safety (FOS) of 1.2.

Fig. 1. Deep excavation for the construction of the basement

In this paper, the design principle is first explained. Based on the measured deflections of the inclinometers installed in the diaphragm walls and JGCs, the performance of the sloping berms serving as a support to the diaphragm wall in an excavation and the effectiveness of the JGCs in improving the sloping berms are then examined. The actual movements of the system are finally compared with the design predictions.

2 Basic Design Concepts

According to Garassino (1997), there is no unique proven calculation procedure to be followed for jet grouting. Many procedures have been adopted, each one with theoretical fundamental and a lot of empirical contents. As a result, the design is still left to the designer's feeling with 99% of empirical background coming from each own personal experience. Projects of the like are the stabilization of an excavated slope (Myers et al. 2003, Corko et al. 2004, Thompson et al. 2009, Ellis et al. 2010) or an embankment (Navin 2005, Kitazume 2008, Filz 2012).

The analysis of JGCs to stabilize a sloping berm to support a diaphragm wall in an excavation involves a very complicated interaction of the diaphragm wall, slope and the JGCs. Morphologically, the JGCs may be perceived as piles pinning on slope such that the JGCs will predominantly provide a horizontal restraint or dowelling to the potentially unstable mass of the slope (Ellis et al. 2010). The JGCs are therefore required to have sufficient penetration into the competent soil layer to transfer the slip force from the weak soil layer to be retained. As such, the design may become more cost effective as the size and spacing of the JGCs increase but there is also increasing risk that the soil will flow through the gap between the JGCs, rather than arching across it. Indifferently, EuroSoilStab (2002) classified jet grouting as one of the deep mixing methods, which aims to produce a stabilized soil mass. The grouting material may interact with the natural soil and not to produce too stiffly stabilized soil mass like a rigid pile which may independently carry out the design load. Thus, the design load should be distributed and carried out partly by the natural soil and partly by the stabilized soil mass (the JGCs). Kitazume (2008) investigated the external and internal stabilities of a grout column type of improved ground under the embankment loading based on a series of centrifugal model tests and elasto-plastic finite element model analysis. For the external stability, a collapse failure pattern in which the deep mixing columns tilt like dominos could take place instead of a sliding failure when the column strength is relatively high. The common design method, which does not take into account this failure pattern, might overestimate the external stability. For the internal stability, the deep mixing column shows various failure modes; shear, bending and tensile failure, depending not only on the ground and external loading conditions but also on the location of each column. However, the common design does not incorporate the effect of these failure modes but only the shear failure mode. Saurer et al. (2011) optimized the JGC spacing by taking into account of the probability distribution of input parameters and based on the Monte Carlo method to calculate the distribution of the global properties of the improved soils. Buschmeier and Frederic (2012) classified JGC as rigid inclusions for ground improvement. Indifferent to the soft inclusions, the JGC material displays a significant permanent cohesion and the stiffness of the inclusion is much larger than that of the surrounding soil, thereby attracting a larger portion of the applied loads. Instead of truly improving the soil, the rigid inclusion acts as a reinforcement of the soil mass. Similarly, Han (2013) classified JGC as semi-rigid columns for ground improvement, and presented the effect of stress concentration ratio to strains in the load transfer mechanism and the failure modes under the embankment loading.

3 Design of a Jet Grouted Sloping Berm to Support the Diaphragm Wall

In this project, several rows of 2 m diameter JGCs at 4 m c/c rectangular spacing were designed as piles on slope (Ellis et al. 2010) initially; each passing through the marine deposits and alluvium with 2 m into the fill above and 2 m into completely decomposed granite (CDG) below. During the course of trial installation to determine the operation parameters, debris such as artificial materials, soil lumps and cobbles from the upper fill layer and inadequate treatment in the CDG were frequently observed in the proof cores. The design was therefore revised to a shorter penetration with 0.5 m nominal embedment into CDG to reflect the quality achieved. Alternatively, use of smaller diameter JGC or better control of the operation parameters to deliver higher energy than in the weaker soils can be used to improve the grouting quality in the CDG but this approach was not adopted because of higher costs. Conventional design principles and design parameters following the local codes and practices (GEO 2000, GEO 2011) to a standard with a FOS equal to 1.2 was adopted.

As shown in Fig. 2, the excavation was carried out in 3 stages: Stage 1 is from the existing ground level of 5.5 mPD to –24.0 mPD in slopes. Stage 2 is further excavation to the formation level of –32.6 mPD in slopes. Stage 3 is the final cutting back of the slopes to the walls. The analysis of two critical excavation stages, i.e. Stage 1 and Stage 2, along grid S33 is presented for illustration.

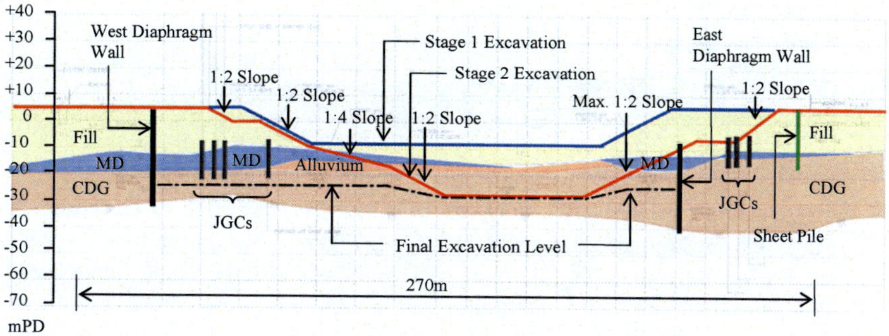

Fig. 2. Geological and excavation profiles along grid S33

The geotechnical design parameters of different soil types are summarized in Table 2. For fine-grained soils, the Young's modulus under undrained condition (E_u) was derived as $E_u = f_u \times S_u$ where S_u is the shear strength and $f_u = 400$. For coarse-grained soils, the Young's modulus (E_s) was obtained as $E_s = f \times N$ where f is the correlation factor equal to 1 and 2 for alluvial sand and CDG respectively and N is the SPT of the soil.

Wall movement around the bermed excavation was predicted using values of elastic modulus equal to two times the modulus of soil (E), i.e. 2E, which is normally accepted for the deformation calculations in Hong Kong. A two-dimensional slope model PLAXIS analysis was then carried out. The JGC slope was modelled as a non-porous elastic perfectly plastic columns with maximum shear strength governed by the Mohr-Coulomb failure criterion and using an interface parameter $R_{int} = 1.0$. Based on the test results of the trial JGCs in the determination of the operation parameters, the unconfined compression strength (UCS) and Young's modulus of JGCs (E_j) were chosen to be 2 MPa and 300 MPa respectively. Because the jet grouted material is quasi-brittle, the shear strain was limited to 0.5%. The external and internal stabilities of the JGC sloping berm

Table 2. Summary of geotechnical design parameters

Formation	Soil Type	Bulk weight (γ) kN/m^3	Friction angle (ϕ)	Cohesion (c) kPa	Young's modulus (E_u or E_s) GPa	Poisson's ratio (ν)
Fill	Clay	18.4	25	1	8,000	0.3
	Sand	19.0	36	0	8,000–20,000	0.3
Marine deposits	Clay	17.2	25	2	12,000	0.3
	Sand	19.0	30	0	14,000	0.3
Alluvium	Clay	19.2	28	2	20,000	0.3
	Sand	19.5	36	0	22,500	0.3
CDG	Above -30mPD	19.0	35	5	30,000–105,000	0.3
	Below -30mPD	19.0	36	3	105,000–200,000	0.3

(Kitazume 2008, Han 2013) were also examined. As the east wall is more critical than the west wall in terms of FOS, only the failure mechanisms of the east wall were obtained by ϕ-c reduction.

The bulk excavation was carried out in three stages; namely, Stage 1 from March 2012 to mid-October 2012, Stage 2 from mid-October 2012 to December 2013 and Stage 3 from January 2014 to March 2014. Figure 3 shows the total displacement of the JGC stabilized sloping berm at east wall for the Stage 1 and Stage 2 excavation. When the JGCs are intact, the failure mechanism of the slope appears like a slip circle, likely because the MDs along the section are quite sandy (see Fig. 3(a) and 3(b)). Failure mechanism by sliding at the bottom of the marine clay is observed in location further north (at grid S23). As this portion of the station structure was later modified to bottom-up construction due to interface problem with the adjacent contractor, further study is therefore not required.

Furthermore, in view of the observed minor cracks in some full depth cores and the probable failure of JGCs by bending, sensitivity analyses assuming the JGCs as frictional materials were therefore carried out. In the model analyses, cracks/fractured materials are hypothetically added to a row of JGC while the others remain intact at the top and frictional at the bottom. Based on the shear box tests, friction angle of the cracked JGC material of 38° is the best-estimate while friction angle of 35° is taken as the worst credible condition. The failure mechanism is found remaining like a slip circle, but is deeper sit into alluvium and CDG in the Stage 2 excavation (see Fig. 3(c) and 3(d)). This appears to show that even when the JGCs are extensively cracked, the JGC would not fail by rotation about its point of embedment, which is explainable because the embedment into CDG is 0.5 m only.

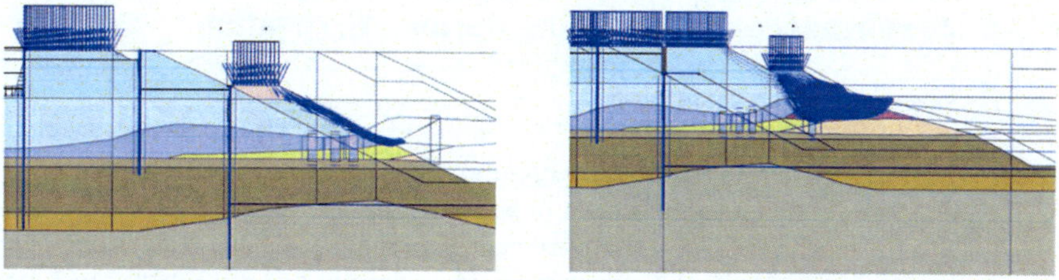

(a) Stage 1 Excavation with Intact JGC

(b) Stage 2 Excavation with Intact JGC

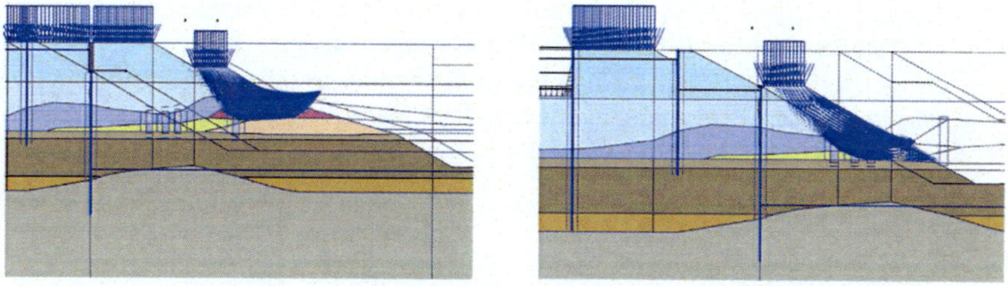

(c) Stage 1 Excavation (with Cracked Parameter $\phi = 35^{\circ}$)

(d) Stage 2 Excavation (with Cracked Parameter $\phi = 35^{\circ}$)

Fig. 3. Probable failure mechanism of JGC treated slope at east wall

Table 3 is a summary of the FOSs of the untreated slope and the JGC stabilized slope at east wall at 2 critical excavation stages. Analytic results show that even without the JGC treatment, the slope is still stable with FOS equal to 1.090 and 1.326 for the Stage 1 and Stage 2 excavation respectively. When JGCs are installed, the FOS will increase to 1.249 and 1.473 for the Stage 1 and Stage 2 excavation respectively. However, if a cracked parameter ϕ equal to 35° is included, the FOS will be reduced by about 9% which may be due to the effective redistribution of the forces among other intact JGCs. The above analysis is also repeated using the actual strength and stiffness of the diaphragm wall, and the FOS is found to be insensitive to these parameters. The JGCs are also analyzed horizontally using PLAXIS to ensure that the soil will not flow through the gap between JGCs, rather than arching across it.

Table 3. FOS of untreated and JGC treated slopes at east wall

Design cases	Stage 1	Stage 2
Case 1: untreated slope without JGC	1.090	1.326
Case 2: treated slope by JGC with Cu = 1000kPa	1.249	1.473
Case 3: treated slope by JGC with cracked parameter $\phi = 35^{\circ}$	1.141	1.365

4 Performance of the Sloping Bermed Excavation

Instrumentation and Monitoring

It is a statutory requirement to provide a monitoring plan where the construction works may affect any nearby building, structure, land, street or services. The geotechnical monitoring of this project has been presented in another paper of the author (So et. al. 2013). Instruments such as inclinometers in diaphragm walls, JGCs and slopes, piezometers in front of and behind diaphragm walls, settlement and movement check points on diaphragm wall, slabs and slopes were installed. Conventional triple-A trigger system is used. The Alert, Action and Alarm levels are assigned as 50%, 80% and 100% of the design predictions. The instruments are grouped and interpreted to serve two purposes.

The first purpose is for the protection of the external sensitive receivers. They are categorized into types for different response actions as different sensitive receivers can tolerate different extent of settlement and angular distortion as seen in Table 4.

Table 4. Categories of sensitive receivers and response actions before reaching alarm level

Sensitive receivers		Response actions before reaching alarm level	
Categories	Examples	Settlement	Angular distortion
Category A (important structures)	Building/structures on piles and seawall	Investigation and mitigation neasures	Investigation and mitigation measures
Category B (semi-flexible utilities)	Gas main, water main, storm water main and sewer	Mitigation measures (and investigation if at junction to buildings)	Investigation and mitigation measures
Category C (flexible utilities and pavements)	Cables	No action unless at junction to buildings	No action unless the change is abrupt
	Carriageways and footpath	Investigation and resurface if the change is abrupt	Investigation and resurface if the change is abrupt

The second purpose is an observational approach to verify the geotechnical design assumptions and monitor the performance of the design at critical stages. Sets of response values are determined from the predicted movements of the systems at various critical stages. Failure mechanisms are identified from the geological models and PLAXIS analysis. Mitigation and contingency measures are considered in advance for prompt actions.

Figure 4 is an abstract of instrumentation layout plan which shows the locations of inclinometers only for clarity. To make the inclinometer measurements, an inclinometer casing is installed in the borehole for ground and JGC or a void former inside a diaphragm wall panel, and the annulus between the ground, JGC and the concrete panel is filled with a bentonite-cement grout. The inclinometer casing has four orthogonal grooves along the inside and these are used to guide an inclinometer probe along in order to survey the line of the inclinometer casing. The inclinometer probe contains orthogonal sensors that measure the angle that the body of the probe makes with vertical. Inclinometer probes also have sprung wheels to hold them in the centre of the inclinometer casing as the survey is taking place. The sum of the angles in each direction enables the shape of the inclinometer casing to be determined and the difference between the shape of a casing at any particular time and the shape of the same casing at a reference time represents the displacement that occurred between the two times.

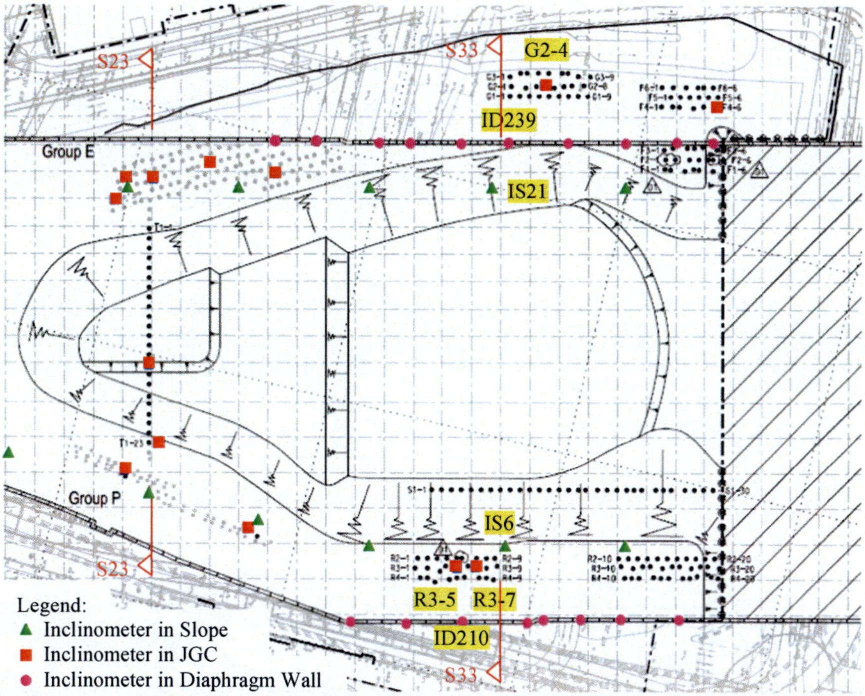

Fig. 4. Layout of the JGCs and location of the inclinometers

In the following section, the east wall and west wall movements are examined based on the measured readings of inclinometers ID239 and ID210 installed in the diaphragm wall panels, and inclinometers G2-4, R3-5 and R3-7 installed in the JGCs. Adjacent inclinometers IS21 and IS6 installed in slope cannot be examined as they were damaged during the course of excavation.

Study of the East Wall Movements

The use of elastic modulus equal to 2E for the deformation calculation is commonly accepted in Hong Kong and used in this paper initially where E is equal to $E_u = f_u \times S_u$ for fine grained soils or $E_s = f \times N$ for coarse grained soils. The bulk excavation started in January 2012 (see Figs. 1 and 2). As shown in Fig. 4(a), inclinometer ID239 showed that the east wall deflected rapidly initially but became steady as the Stage 1 excavation to

–24.0mPD progressed and completed, but is close to the predicted movement of 108 mm for the whole excavation already. The predicted movement using 2E was reviewed by an expert panel and was considered more appropriate to use 1E in this bermed excavation which is to be discussed in later. The Stage 2 excavation started in mid-October 2012. Likewise, the east wall deflected rapidly initially and became steady as the excavation progressed to –32.4 mPD and completed. Indeed, Figs. 4(a) and 4(b) show that the deformed profile of the wall in September 2013, when the Stage 2 excavation is about

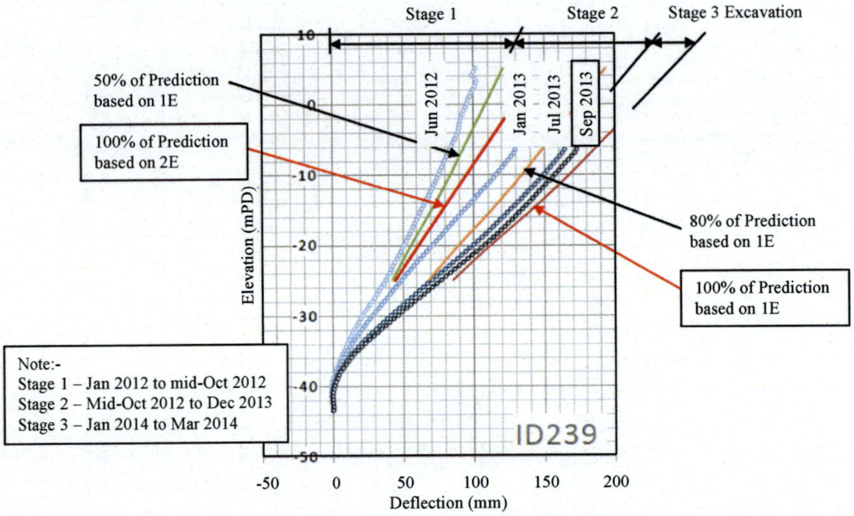

(a) – Deflected Profiles of Diaphragm Wall

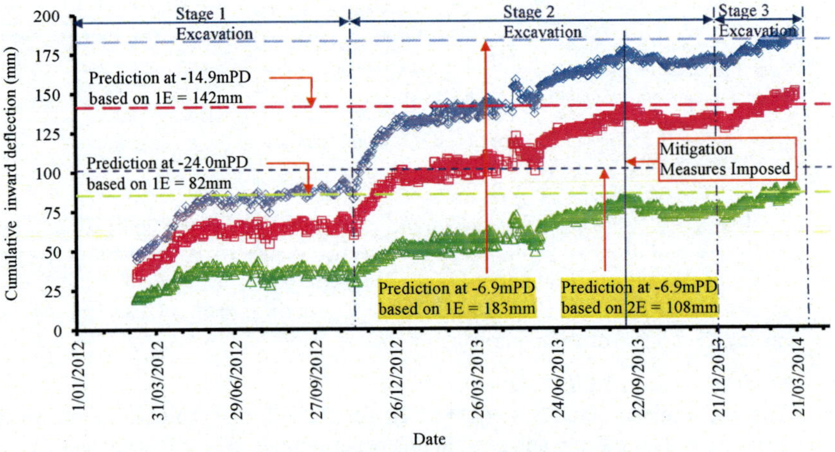

(b) – Measured East Wall Movement during Excavation

Fig. 5. East wall movement during excavation

90% completed, has largely exceeded the deflection predicted using 2E and close to the prediction using 1E. Mitigation measures are imposed such as reducing the construction load at the back of the wall and some recovery was achieved. However, with further excavation towards the diaphragm wall face in the Stage 3 excavation commenced in January 2014, the wall movement continued to increase rapidly even with slight increase in the excavated volume. All excavation was completed in March 2014 and the final movement slightly exceeded the prediction using 1E.

The unexpected movement was explained by Burland (2012) in his expert review report that: "When the material in the base of the excavation is unloaded, experience shows that the stiffness of such an overconsolidated material under changes in the vertical load are greater than for the equivalent normally consolidated material. Hence, the value of elastic modulus adopted for the analysis is normally taken as twice the normally consolidated values, i.e. 2E. However, this is reasonable in an excavation when it is only dealing with reversals in vertical loading and vertical strain, but may not be the case in a sloping berm subjected to changes in both horizontal loading and to shear stresses as shown in Fig. 5. Indeed, the Plaxis analyses for failure mechanisms confirm that the mode of straining within the final form of the slopes is predominantly by shearing rather than by direct straining (Fig. 3). The deformations are therefore controlled largely by the values of shear modulus G within the slopes. There is no reason to believe that the reductions in horizontal stress and the increases in shear stress within the slopes during excavation will result in increases in the values of G. Indeed the reverse is probably true due to the non-linear nature of the stress-strain relationships. In view of the above, it is not surprised that the use of 2E underestimated the predicted movements" since $G = E/(2(1 + \nu))$ where ν is the poison ratio.

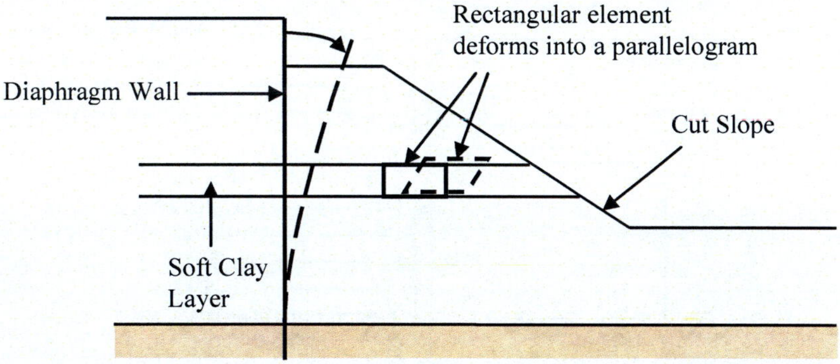

Fig. 6. Diagram showing mode of deformation with soft clay layer (Burland 2012)

The underestimation of wall movement when using 2E, seems explainable from the findings of Kumruzzaman and Yin (2010) as well. They observed that the rotation of principal stress directions are common in many geotechnical engineering problems and may have significant influences on the behaviour of soils. Based on a series of undrained tests on compacted hollow cylinder specimens of CDG in hollow cylinder apparatus, they observed that the deviator stresses and the excess pore pressure decrease with the vertical angle, and the principal stress direction angle has a significant influence on the strength parameters.

Figure 6 shows the movements of the JGC located at the back of the diaphragm wall and the surrounding soils measured by inclinometer G2-4 during bulk excavation before it was damaged in April 2013. The JGC is about 12 m long. Deformed profiles show that the JGC had very little movement during Stage 1 excavation, seemingly implying that the effectiveness of the dowel action. However, the JGC started to translate in the MDs and alluvium, but slipped in the CDG when Stage 2 excavation started The movement was very quick in November 2012 to December 2012 but lesser in magnitude as the excavation proceeded which coincides with the time when the wall movements increased (Figs. 4(a) and 4(b)). The sliding of the JGC in the CDG is consistent with the PLAXIS study of probable failure mechanisms and appears to indicate the JGC "flowed" with the soil mass when it slipped, like the sliding mode of failure of the JGC in Kitazume (2008). The movement of the JGC and the diaphragm wall are expected to be reduced if the JGC is intact and penetrates deeper into the CDG to intercept the slip surface such that a larger horizontal restraint or better dowelling to the mass of the slope can be provided.

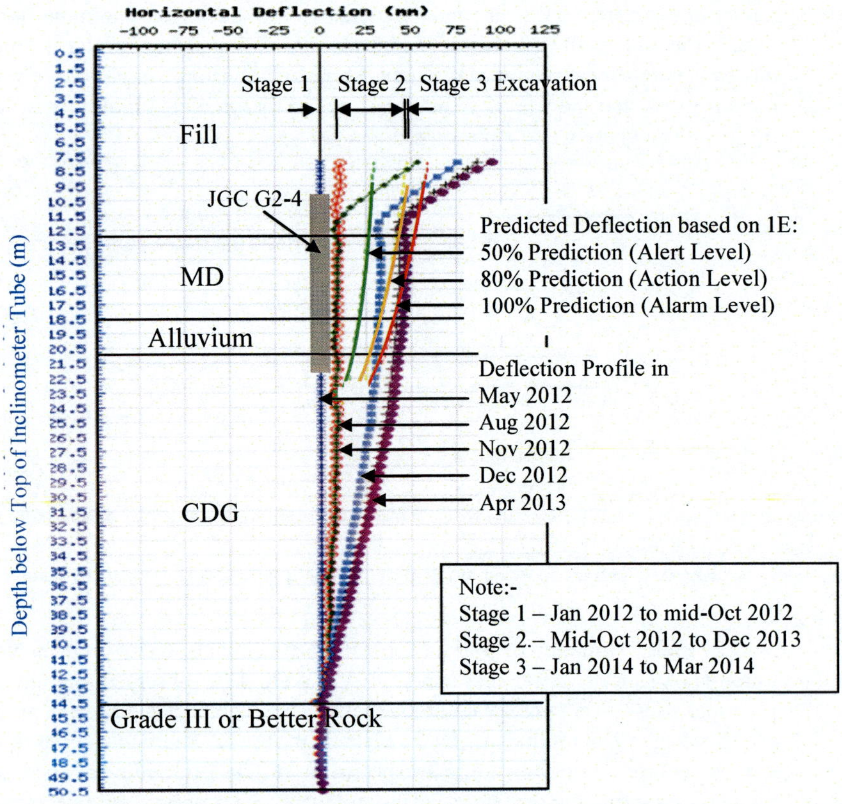

Fig. 7. Deflection profiles of inclinometer G2–4 in JGC

Study of West Wall Movements

Figure 7 shows the movements of the west diaphragm wall measured by inclinometer ID210 during bulk excavation. Indifferent to the east wall, the actual movements of the diaphragm wall occurred mostly in the Stage 1 excavation, then lesser in magnitude in the Stage 2 excavation and very little in the Stage 3 excavation. The actual movements are found close to the predictions using 2E because the berm is wide as the middle portion is less steep and the mode of straining of the soil within the slope is predominantly by direct straining and not by direct shearing. Figure 8 shows the movements of the JGCs in front of the diaphragm wall by the inclinometers R3–5 and R3–7. The JGCs are about 10 m long. They showed similar movements as the wall at the early stage of the excavation, and appears to rotate with the slope movement, indicating probable rotational mode of failure of the JGC (Kitazume 2008). The movement of the JGCs and the diaphragm wall are expected to be reduced if the JGC is intact and penetrates deeper into the CDG passing through the "centre" of rotation such that a larger horizontal restraint or better dowelling to the mass of the slope can be provided.

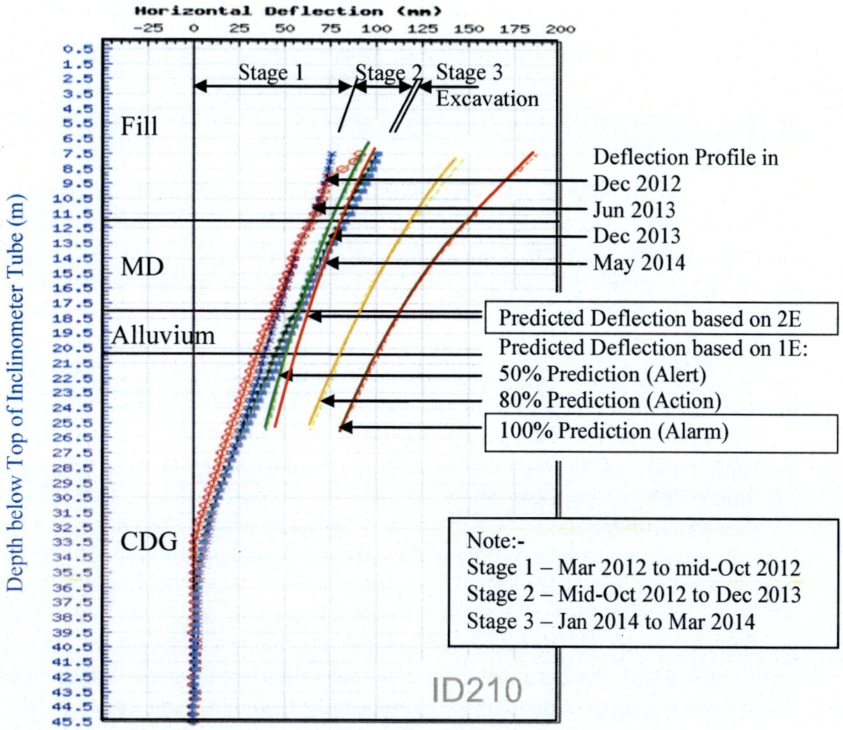

Fig. 8. Deflection profiles of inclinometer ID210 in diaphragm wall

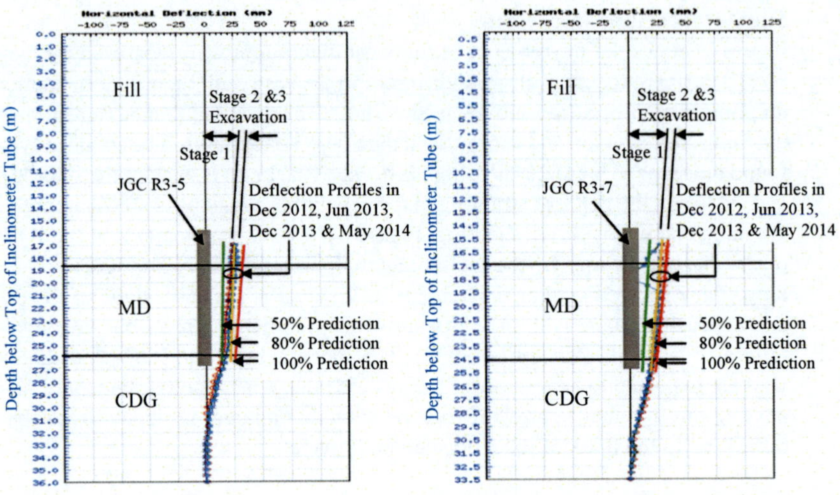

(a) Defection Profiles of Inclinometer R3-5 in JGC (b) Deflection Profiles of Inclinometer R3-7 in JGC

Fig. 9. Deflection profiles of inclinometers R3-5 and R3-7 in JGC

5 Conclusions

The use of jet grouting to stabilize a sloping bermed excavation is unconventional and unprecedented. It appears that there is no unique proved design method to follow. Some designers may perceive the JGCs as piles on slope while others may consider the JGCs as soil stabilization works. In this project, the concept of piles on slope was adopted. The external and internal stabilities of the system were checked using conventional design principles and parameters, following the local codes and practices. In view of the debris and untreatment in CDG frequently observed in the proof cores, the design was revised from 2.0 m penetration to 0.5 m nominal embedment into CDG to reflect the quality achieved. As minor cracks were observed in some full depth cores which may result probable failure of the JGCs by bending, sensitivity analyses assuming the JGCs as frictional materials were carried out.

The east wall has the JGCs behind the wall. Inclinometer in diaphragm wall showed that the actual movement is larger than the conventional analysis of the movement around an excavation using 2E. This is consistent with the model findings that the mode of straining within the final form of the slopes is predominantly by shearing rather than by direct straining. The deformations are therefore controlled largely by the values of G within the slopes. Furthermore, the reductions in horizontal stress and the increases in shear stress within the slopes during excavation will not result in increases in the value of G, but is due to the non-linear nature of the stress-strain relationships. It is therefore not surprised that the use of 2E would underestimate the predicted movements. Inclinometer in JGC showed that the JGCs behind the wall translated in the marine deposits and alluvium but slipped in the CDG when the wall movements increased. The sliding of the JGC in CDG appears to indicate that the JGC "flowed" with the soil mass when it slipped like the sliding mode of failure of JGC. This implies that the movement of the diaphragm wall can be reduced if the JGCs are larger in size, closer in spacing or deeper in penetration into the CDG such that a larger horizontal restraint to the slope mass can be provided.

The west wall has the JGCs in front of the wall. Indifferently, inclinometer in diaphragm wall showed that the actual movement is close to the predictions using 2E. This is likely because the berm is wide and the mode of straining of the soil within the slope is predominantly by direct straining rather than by shearing. Inclinometers in JGCs showed that the JGCs in front of the wall appeared to rotate with the slope movement indicating probable rotational mode of failure of the JGC. Likewise, this implies that the movement of the diaphragm wall can be reduced if the JGCs are larger in size, closer in space, or deeper in penetration into the CDG such that better dowelling to the slope mass can be provided.

The above observed performance demonstrates the importance of, not only the checking of the internal and external stability of the system, but also the understanding of the probable failure mechanisms and monitoring of the performance. These help the designer to understand the design intents of the JGCs (in compression, shear, bending or their combinations) so as to determine and optimize the quality requirements, embedment length, size, spacing and arrangement (in single, buttress or block).

References

Akan, R., Keskin, S.N., Uzundurukan, S.: Multiple regression model for the prediction of unconfined compressive strength of jet grout columns. In: Procedia Earth and Planetary Science 15(2015), World Multidisciplinary Earth Sciences Symposium, WMESS, pp. 299–303. Elsevier (2015)

Atangana, N.P.G., Chen, J., Modoni, G., Arulrajah, A., Kim, Y.-H.: A review of jet grouting practice and development. Arab. J. Geosci. **11**(16), 1–31 (2018). https://doi.org/10.1007/s12517-018-3809-7

Atangana, N.P.G., Shen, J.S., Modoni, G., Arulrajah, A.: Recent advances in horizontal jet grouting (HJG): an overview. Arab J. Sci. Eng. **43**(4), 1543–1560 (2017)

Bayesteh, H., Sabermahani, M.: Field study on performance of jet grouting in low water content clay. Eng. Geol. 264, 105314 (2020). Elsevier

Bearce, R.G., Mooney, M.A., Kessoun, P.: Estimation of jet grout column geometry using a DC electrical resistivity push probe. In: International Symposium on Non-Destructive Testing in Civil Engineering, Berlin, Germany, 15–17 September, 2015 (2015)

Belleto, D., Schon, J., Spagnoli, G.: Mathematical analysis of shadown effect in jet grouting. J. Geotech. Geoenvironmental Eng. ASCE. **144**(12), 04018088 (2018)

Brill, G.T., Burke G.K., Ringen, A.R.: A ten-year perspective of jet grouting: advancements in applications and technology. In: Geotechnical Special Publication No. 120, Grouting and Ground Treatment, Proceedings of the 3rd International Conference, New Orleans, Louisiana. pp. 218–235 (2003)

Burke, G.K., Meffe, D.A.: Fixing foundations. Civ. Eng. ASCE. **61**(3), 63–65 (1991)

Burke, G.K.: The state of practice of jet grouting. Grouting and Deep Mixing 2012, Geotechnical Special Publication No. 124, ASCE, Reston, pp. 875–886 (2012)

Burland, J.: Interim expert review panel - review meeting report for the mass transit railway corporation projects. Review Report (2012)

Buschmeier, B., Masse, F.: Discussion of differences in design technology between granular and grouted inclusions. Menard Ground Improvement Specialists, XXVI Reunion Nacional de Mecanica de Suelos e Ingenieria Geotecnica, Cancun, Mexico (2012)

Cheng, S.H., Liao, H.J., Yamazaki, J., Wong, R.K.N.: Evaluation of jet grout column diameters by acoustic monitoring. Can. Geotech. J. **54**(2), 1781–1789 (2017)

Corko, D., Maric, B., Lovrencic, D., Tomac, I.: Application of jet grouting in slide remediation. In: International Congress on Soil Improvements in Place, ASEP-GI 2004, Paris, 9–10 September 2004 (2004)

Croce, P., Flora, A., Mondoni, G.: Jet Grouting: Technology, Design and Control, pp. 9–25. Taylor and Francis Group, Boca Raton (2014)

Ellis, E.A., Durrani, I.K., Reddish, D.J.: Numerical modelling of discrete pile rows for slope stability and generic guidance for design. Geotechnique **60**(3), 185–195 (2010)

EuroSoilStab. Development of design and construction methods to stabilize soft organic soils: design guide for soft soil stabilization. CT97–0351, European Commission, Industrial and Materials Technologies Programme (Rite-EuRam III), Bryssel (2002)

Filz, G.M.: Load transfer, settlement, and stability of embankments founded on columns installed by deep mixing methods. In: A Geotechnical Engineering Seminar Presentation, National Technical University of Athens (2012)

Garassino, A.L.: Design procedures for jet-grouting. In: Seminar on Jet Grouting organized by The Foundation Equipment Pte Ltd and Gaggiotti Far East Pte Ltd, Singapore (1997)

GEO. Guide to retaining wall design (Geoguide 1). Geotechnical Engineering Office, Hong Kong Special Administrative Region (2000)

GEO. Geotechnical manual for slopes. Geotechnical Engineering Office, Hong Kong Special Administrative Region (2011)

Han, J.: Recent advances in column technologies to improve soft foundations. In: Annual Kansas City Specialty Geotechnical Seminar sponsored jointly by ASCE Kansas City Section – Geotechnical Section, Association of Engineering Geologists, Kansas City – Omaha Section and University of Missouri at Kansas City, Departs of Geosciences and Civil Engineering (2013)

Ho, C.E.: Turbulent fluid jet excavation in cohesive soil with particular application to jet grouting. PhD Thesis, Massachusetts Institute of Technology (2005)

Kitazume, M.: Stability of group column type DM improved ground under embankment loading behaviour of sheet pile quay wall. Report of the Port and Airport Research Institute, Report vol. 47, no. 1 (2008)

Kumruzzaman, M., Yin, J.H.: Stress-strain behaviour of completely decomposed granite in both triaxial and plane strain conditions. Jordan J. Civ. Eng. **6**(1), 83–108 (2010)

Lin, C.P., Lin, C.H., Ngui, Y.J., Wu, P.L.: Jet grouting column diameter measurement using in-hole electrical resistivity tomography. In: The 5th International Conference on Engineering Geophysics, Al Ain, UAE, 21–24 October, 2019 (2019)

Meyers, J., Myers, T., Petrasic, K.: Jet grout stabilization of steeply excavated soil slope. In: Proceedings of the 3rd International Conference, New Orleans, Louisiana, vol. 1, pp. 318–329 (2003)

Modoni, G., Bzowka, J.: Analysis of foundation reinforced with jet grouting. J. Geotech. Geoenvironmental Eng. **138**(12), 1442–1454 (2012)

Navin, M.P.: Stability of embankments founded on soft soil improved with deep-mixing-method columns. PhD Thesis, Virginia Polytechnic Institute and State University (2005)

Ribeiro, D., Cardoso, R.: A review on models for the prediction of the diameter of jet grouting columns. Eur. J. Environ. Civ. Eng. **21**(6), 1–29 (2017)

Saurer, E., Marcher, T., Lesnik, M.: Grid space optimization of jet grouting columns. In: Proceedings of the 15th European Conference on Soil Mechanics and Geotechnical Engineering, pp. 1055–1060 (2011)

Shen, S.L., Wang, Z.F., Ho, C.E.: Current state of the art in jet grouting for stabilizing soft soil. In: Ground Improvement and Geosynthetics, GPS 238, pp. 107–116. ASCE (2014)

Shibazaki, M.: State of practice of jet grouting. In: Geotechnical Special Publication No. 120, Grouting and Ground Treatment, Proceedings of the 3rd International Conference, New Orleans, Louisiana. pp. 198–217 (2003)

So, A.K.O., Ko, P.W.L., Man, V.K.W.: Geotechnical instrumentation monitoring for the construction of the west kowloon terminus of the express rail link. In: The 4th International Symposium on Geotechnical Safety and Risk, 4–6 December, 2013, Hong Kong (2013)

Thompson, W.R., Jeffrey, R.H., Loehr, J.E.: Case history: value engineering of driven h-piles for slope stability on the Missouri river. In: Proceedings of International Foundation Congress and Equipment Expo 2009, Orlando, Florida, pp. 207–214 (2009)

Toraldo, C., Modoni, G., Ochmanski, M., Croce, P.: The characteristic strength of jet grouted material. Geotechnique **68**(3), 262–279 (2017)

van der Stoel, A.E.C.: Grouting for pile foundation improvement. PhD Thesis, Delft University of Technology, the Netherlands (2001)

Wang, Z.F., Shen, S.L., Ho, C.E.: Jet grouting practice: an overview. Geotech. Eng. J. SEAGS AGSSEA. **44**(4), 88–96 (2013)

Wang, Z.F., Shen, S.L., Ho, C.E., Xu, Y.S.: Jet grouting for mitigation of installation disturbance. Proc. Inst. Civ. Eng. Geotech. Eng. **167**(GE6), 526–536 (2014)

Wanik, L., Mascolo, M.C., Bzowka, J., Modoni, G., Shen, S.L.: Experimental evidence on the strength of soil treated with single and double fluid jet grouting. In: Proceedings of the Grouting: Grouting, Deep Mixing and Diaphragm Walls, Honolulu, 9–12 July, 2017 (2017)

Wu, Y.D., Diao, H.G., Ng, C.W.W., Liu, J.: Investigation of ground heave due to jet grouting in soft clay. Technical Note, Journal of Performance of Constructed Facilities, ASCE (2016)

Yoshida, H.: Recent developments in jet grouting. In: Proceedings of the 4th International Conference on Grouting and Deep Mixing, New Orleans, Louisiana, United States, pp. 1548–1561 (2012)

Zohrer, A.: innovative design for retaining structures using combined products. Geotehnika, e-casopis Drustua za geotehnika u Bosni I Hercegovini. (2017). ISSN2303–8403, Broj 3

Influence of Lime and Coal Gangue on the CBR Behavior of Expansive Soil

Mohammed Ashfaq(✉) and Arif Ali Baig Moghal

Department of Civil Engineering, National Institute of Technology, Warangal 506004, Telangana, India

Abstract. The consistent surge in the utilization of coal has resulted in its higher production. Coal gangue is a by-product in the coal mining process with a wide range of application. Understanding the geotechnical behavior of coal gangue (CG) and its interaction with sensitive and problematic soils provides viable solutions for its large-scale utilization. Bulk utilization of CG can be attained if it has the potential for subgrade material. In the current study, the effect of CG addition (10%–50% by dry weight of soil) on the California Bearing Ratio (CBR) behavior of expansive black cotton (BC) soil is studied. Further, lime (2%, 4% and 6%) as an additive is considered due to its cementitious properties. CBR tests were conducted on both soaked and unsoaked conditions. The Tangent Modulus(TM) and Secant Modulus (SM) were evaluated from the stress (load)-strain (penetration) curves to understand the stiffness characteristics of the CG-BC soil mixture. Further, attempts were made to estimate resilient modulus (M_R) from the observed CBR values using existing correlations. The results from the study showed that soaked CBR value of BC soil increased from 4 to 23 with 40% CG, due to better mobilization of frictional resistance. The CBR values decreased beyond 40% due to a reduction in the cohesive component of BC-CG mixture. In the presence of lime, the BC-CG mixtures yielded better soaked and unsoaked CBR values, with 6L and 40CG outperforming all other combinations. The TM, SM and M_R values increased with coal gangue addition, which is proportional to the increase in CBR values.

Keywords: Coal gangue · CBR · Subgrade · Expansive soils · Lime

1 Introduction

Expansive soils undergo rapid volume changes with the seasonal moisture fluctuations. The aggressive volumetric change causes significant damage to overlying structures, especially in highway construction (Petri and Little 2002; Senol et al. 2006). The problems associated with the expansive soil and the high cost of soil replacement has augmented research in the stabilization techniques through physical and chemical alteration. The alteration of the mechanical properties of expansive soil using chemical additives has gained universal acceptance due to its more excellent repeatability and efficiency (Puppala et al. 2003). Traditionally, the calcium-based additives like hydrated

J. Neves et al. (Eds.): GeoChina 2021, SUCI, pp. 102–113, 2021.
https://doi.org/10.1007/978-3-030-80155-7_7

lime and cement were used in the stabilization of expansive soils (Darikandeh 2018). Non-traditional additives in the stabilization technique include solid industrial wastes and combustion residues such as coal gangue, cement kiln dust, red mud and fly ash. The lime addition has proven to be an effective technique to address the swell and shrinkage problem coupled with enhanced strength and workability of expansive soils. The lime usage has also been extended to in-situ stabilization techniques in the form of lime-soil columns and deep soil mixing (Phanikumar 2009).

Coal mining annually generates millions of tons of solid wastes in the form of coal combustion residue (CCR), coal gangue (CG) and coalmine overburden (Ashfaq et al. 2020). The economic and environmental benefits of utilizing CG as fill material in earthworks was established by previous works (Indraratna et al. (1994); Indraratna et al. (2012); Ashfaq et al. (2020a); Ashfaq et al. (2020b)). Expansive soil, in its native form, does not possess the desired California bearing ratio (CBR) values to meet the design specifications of a subgrade/subbase material. To enable expansive soil to be used for road embankments and subgrade of the pavements, understanding the CBR behavior is essential. Thus, the stabilization of expansive soil not only address the problem of swelling/shrinkage but also imparts significant strength (Pandian and Krishna 2003). Earlier studies by Bell (1993), Pandian and Krishna (2002) and Phanikumar (2009) have recognized the enhancement of strength and CBR properties of expansive soil stabilized with fly ash-lime and fly ash-cement mixtures. In an earlier study by Tasalloti et al. (2015), the authors have attempted to use coal gangue as an additive in the stabilization of steel furnace slag to enable its utilization as fill material. However, the attempts to use CG as a sufficient admixture to stabilize expansive soil are sparse. In the present study, the effect of varying quantity of CG addition on the CBR of Black Cotton (BC) soil is studied. Further, lime as an additive was considered due to its potential to react with the clay mineral surface and consequent formation of cementitious products.

Further, the motive for considering CG-lime mixture to stabilize expansive soil is economy (CG being waste residue), workability and repeatability (the wider availability and utilization of lime) in field applications. Furthermore, attempts were also made to identify the optimum additive combination for the stabilization of CG-BC soil mixture. The TM and SM were evaluated to understand the stiffness characteristics of the CG-BC soil mixture. Further, attempts were also made to estimate resilient modulus (M_R) from the observed CBR values using existing correlations.

2 Materials and Methodology

BC soil is locally obtained (from the institute campus) at a depth of 1 m from the surface level. CG samples used in the study were obtained from Kakatiya coalmines, bhupalpally (18° 25′ 53.2″ N; 79° 51′ 30.8″ E), Telangana state, India. The hydrated lime (Taranath Scientific and Chemical Company) was used which is of AR Grade (Analytical Reagent Grade) having purity of more than 99% (Assay content < 1%). The physical properties of coal gangue and BC soil and the corresponding ASTM standards adopted for the tests are presented in Table 1. For each combination, the BC soil is dry mixed (before saturation) with lime and coal gangue based on its observed MDD values. In order to avoid the influence of moisture fluctuations on the CBR behavior, fixed optimum moisture content of 21% (OMC of BC soil) was employed for all the combinations. The lime content of 2%, 4% and 6% by dry weight of soil mixture (coal gangue and BC soil) was adopted based on the earlier studies (Murty and Krishna 2006) on lime stabilization of BC soil (source material). The dry mixing, compaction and load application stages of the CBR test were done instantly to avoid alteration in properties with delay in compaction. The surcharge load of 0.05 kg/cm^2 was applied to the plunger and penetration rate of 1.25 mm/min was maintained for all the tests. In soaked condition, the surcharged load (0.05 kg/cm^2) was applied for all the testing samples before the soaking period of four days. The curing periods of 7 (3 days curing and four days soaking), 14 (10 days curing and four days soaking) and 28 days (24 days curing and four days soaking) were considered in the study. Further, TM and SM values (represent the slopes of initial straight-line portion and the line joining zero to one half of peak stress value respectively) were determined from stress (load) - strain (penetration) curves for all the combinations.

Table 1. Physical properties of Coal Gangue and BC soil

Property	Value		Code
	CG	BC soil	
Liquid limit (%)	28	76	ASTM D4318 (2017)
Plastic limit (%)	NP	22	ASTM D4318 (2017)
Plasticity index (%)	NP	54	ASTM D4318 (2017)
Specific gravity	2.56	2.74	ASTM D854 (2014)
Maximum dry density (g/cm^3)	2.1	1.68	ASTM D698–12e2 (2015)
Optimum moisture content (%)	17	21	ASTM D698–12e2 (2015)
pH	7.24	7.8	ASTM D4972 (2019)
USCS classification	SP	CH	ASTM D2487 (2017)
CBR (%)	9	4	ASTM D1883 (2016)

3 Results and Discussions

The CBR behavior of CG-BC soil mixtures was studied to evaluate the feasibility of its application as a subgrade/subbase material in pavements. From the physical properties of untreated BC soil and CG presented in Table 1. it can be observed that the CG is non-plastic with negligible clay fractions. Further, the specific gravity of CG is lower compared to BC soil, which may be due to the lower iron content of CG (Ashfaq et al. 2020b). The natural pH values of both CG and BC soil confirm their chemically inert behavior. The grain size distribution (GSD) curves of BC soil and CG samples are presented in Fig. 1. It is evident from Fig. 1. that the fines content in BC soil is much higher with more significant clay fraction compared to CG. The variation in the gradation

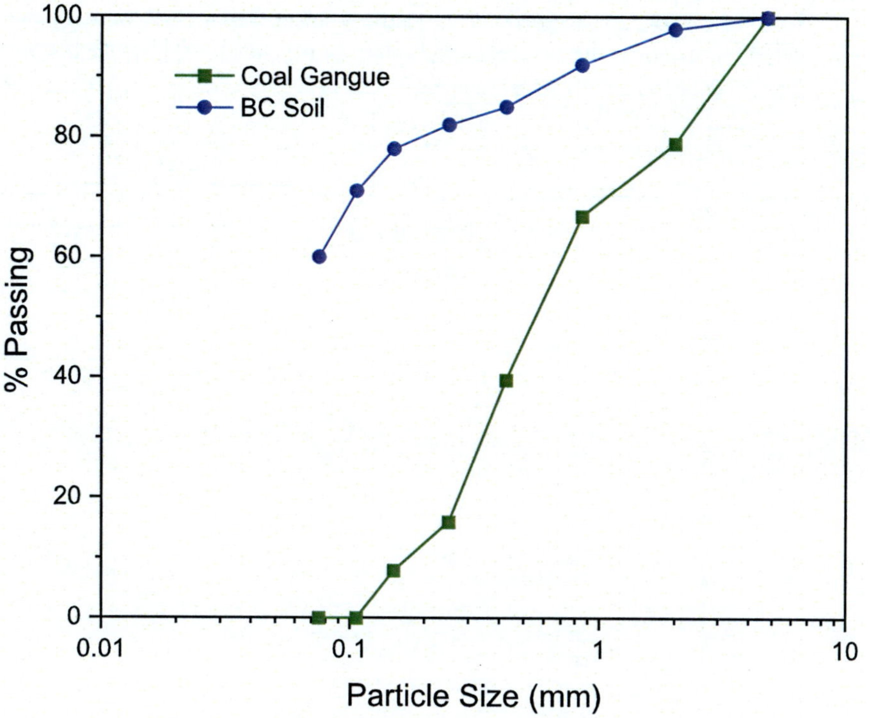

Fig. 1. Grain size distribution curves of coal gangue and BC soil

curve is evident with CG exhibiting steeper curve, which can be attributed to its more significant coarse fraction. The role of coal gangue, lime content and curing period on the CBR of BC soil is presented in the following sections.

3.1 Effect of Coal Gangue

The effect of CG on the CBR of BC soil mixtures are presented in Fig. 2. From the results, it can be noted that the addition of CG has increased CBR values of BC soil. The maximum increase in CBR was noted for 40% CG addition and subsequent CG addition has little or no effect on the CBR of BC soil- CG mixture. The variation in soaked CBR value is found to be identical to the unsoaked condition with the highest value observed for 40% CG addition. However, the sudden spike in CBR values was observed for 20% and 40% CG addition. The first spike (with an increment of 100%) was observed for 20% CG addition which can be attributed to relatively denser packing attained by mixing of coarse fractions of CG with a sufficient fraction of BC soil. The second peak (with an increment of 55%) at 40% CG addition may be due to the apparent cohesion imparted by the BC soil contributing to the more excellent CBR value. Further, the rise in CBR of CG-BC soil mixture can also be attributed to the mobilization of frictional resistance with the addition of CG. The identical peaking pattern was also noted for soaked CBR value of CG-BC soil mixtures. The relatively lower fall in CBR values under soaked condition indicates the minimal loss in bearing resistance with the capillary action of water. Similar observations were made by Pandian et al. (2001) for fly ash stabilized BC soil and Ashfaq et al. (2020) for CG.

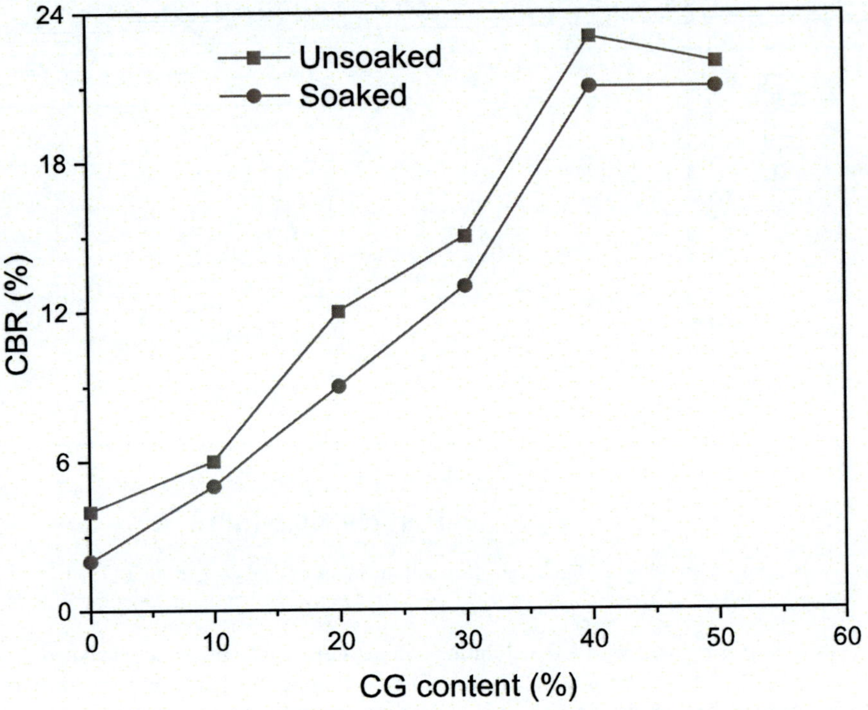

Fig. 2. Effect of CG on CBR values of BC soil

3.2 Effect of Lime Addition

The variation in CBR values (soaked and unsoaked) with lime addition for CG-BC soil mixtures at 28 days curing period are presented in Fig. 3. From the results, it is apparent that the addition of lime has substantially increased the CBR values for all the combinations. The increasing linear pattern with lime addition is observed for both unsoaked and soaked conditions. The highest increment in CBR value was observed at 2% lime addition, and subsequent lime addition has shown relatively lower increment. Inconsistent with the results reported for CG-BC mixtures, the optimum CBR value was noted for 40% CG addition. For unsoaked and soaked conditions, the highest CBR values were reported for 6% lime addition with CBR values of 44.2 and 33.3 respectively. The increase in CBR of CG-BC soil mixtures with lime addition can be attributed to the following three reasons:

i. The addition of lime enhances the pH of the pore solution of the CG-BC soil mixture, which contributes to the dissolution of silica and alumina from both CG and BC soil. The dissolved silica and alumina from clay readily forms primary hydration compounds in the form of calcium silicate and calcium aluminate
ii. Higher lime contents facilitate the formation of a more significant amount of hydration products, thus contributing to the greater CBR values at higher lime content.
iii. The affinity of lime above the lime fixation point (lime dosage consumed by clay minerals to form hydration products) to form hydration compounds with reactive silica and alumina from coal gangue (Bell 1993; Pandian and Krishna 2002).

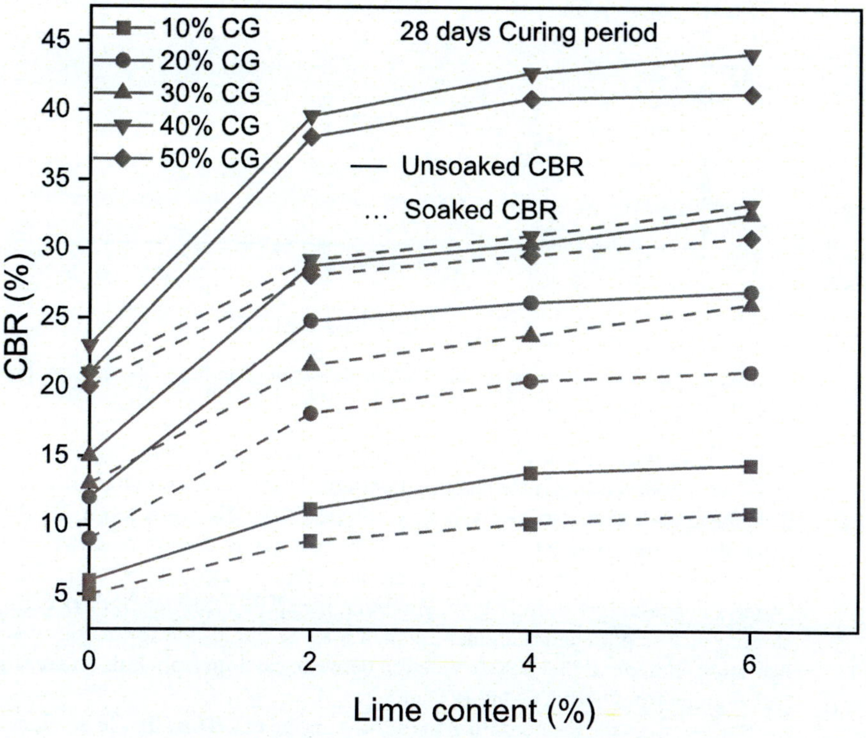

Fig. 3. Effect of lime on CBR of CG- BC soil mixtures

3.3 Effect of Curing Period

The variation in CBR values (unsoaked and soaked) of lime (due to brevity, only results for 6% lime additions are presented) stabilized CG- BC soil mixtures are presented in Fig. 4. The CBR values exhibited increasing pattern with the curing period and accordingly, the highest values were noted for 28 days curing periods with CBR values of 44.2

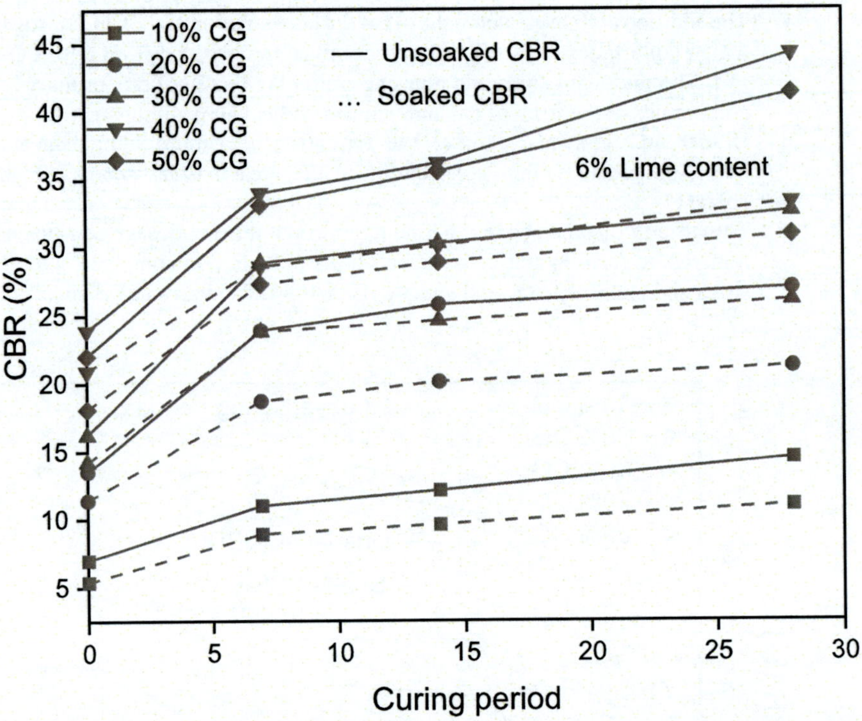

Fig. 4. Effect of Curing period on CBR values of lime stabilized CG- BC soil mixtures

and 33.3 for unsoaked and soaked conditions, respectively. The highest rise in CBR value was observed for an initial curing period of 7 days with an increment of 77% and 67% for unsoaked and soaked conditions respectively. The swift rise in CBR value for seven days curing indicates the formation of primary hydration reactions is complete within the initial curing period of seven days. Accordingly, the rise in CBR value at 14 days is gradual, with little gain. However, the CBR values at 28 days curing period showed more significant increment, which may be due to the formation of secondary hydration compounds because of reaction between calcium from lime and reactive silica and alumina from both CG and BC soil.

Further, from the results shown in Fig. 4, it can be noted that the CBR values of all the combinations drastically decreased with the curing period under soaked condition. This contrasting behavior is due to the loss of surface tension forces upon soaking, which is greater than the gain in strength due to the formation of hydration compounds. Similar observations were made Pandian, and Krishna (2003) and Moghal et al. (2017) for Fly ash and lime stabilized BC soil.

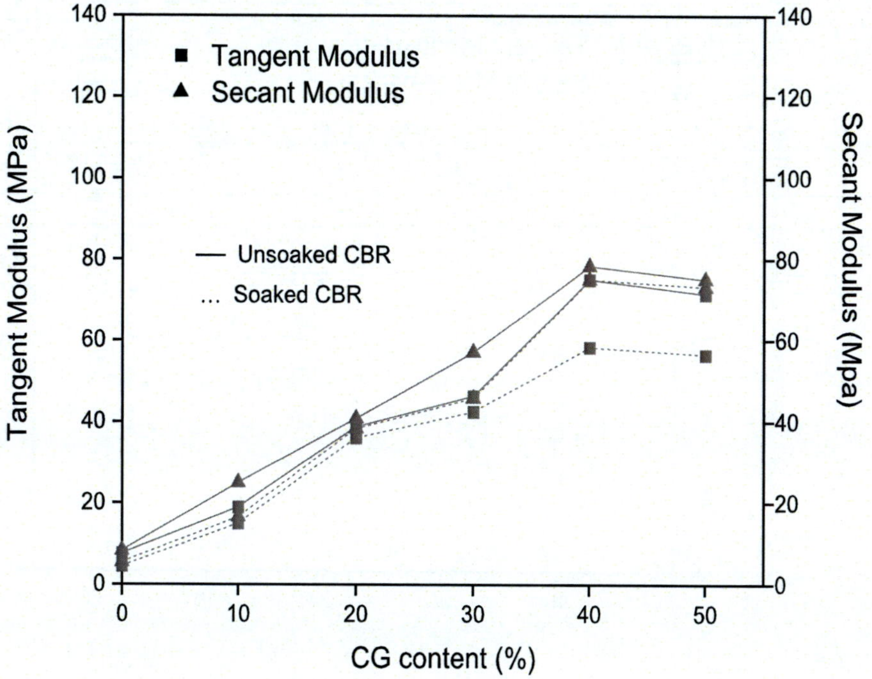

Fig. 5. The variation in TM and SM values of CG-BC soil mixtures

3.4 Effect of Coal Gangue on Elastic Moduli

The elastic moduli reflect the stiffness behavior of soil under static loading condition. The variation in TM and SM with varying coal gangue content for both control and lime treated case is presented in Fig. 5 and Fig. 6, respectively. The TM and SM values for the control case are 7.6 and 8.3 MPa respectively in unsoaked condition, which reduced to 4.5 and 5.5 MPa in soaked condition. The TM and SM values increased with CG content for both control and lime treated case with the highest increment observed for 40% CG addition. The decrease in TM and SM values under soaked condition is consistent for both the cases, with the relatively higher decrease observed for the lime treated case. For control case, relatively marginal rise in TM and SM values was observed. For the lime treated case, the rise is substantially more significant with the highest increment of 540% and 255% in TM and SM respectively observed for 10% coal gangue addition. For the lime treated case, the subsequent addition of coal gangue beyond 10% has caused a gradual increase in both TM and SM values. Further, it was observed that the TM values are relatively lower than the SM for both controlled and lime treated case, which can be attributed to early straightening of the stress-strain curve. Further, it was also observed that the increase in both the elastic moduli is proportional to the increase in CBR values. Thus, the hypothesis presented for increment in CBR values is also valid for the variation the TM and SM variation.

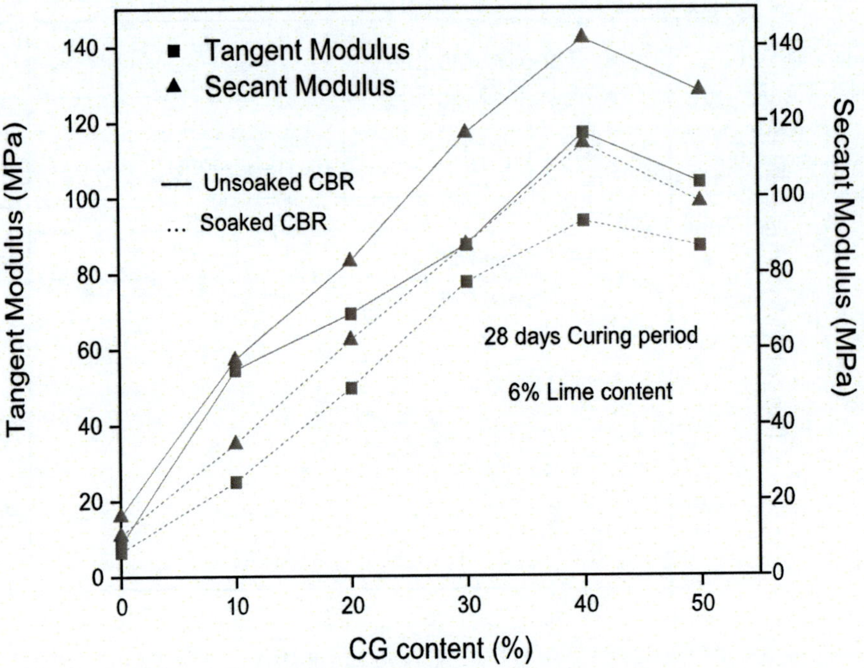

Fig. 6. The variation in TM and SM values of lime stabilized CG-BC soil mixtures

3.5 Correlation Between CBR and Resilient Modulus Values

The resilient modulus (M_R) is one of the essential parameters in the design of pavement structures. Hence, an attempt was made to estimate the M_R values based on the observed CBR values. Previous researchers proposed many empirical correlations to estimate M_R from CBR values which were analyzed (Table 2) to select the best possible model for the estimation of M_R value.

Table 2. The existing correlations for the estimation of resilient modulus based on CBR value.

Correlation model	Reference
$M_R = 10$ CBR	Heukelom and Foster (1960)
$M_R = 38$ CBR$^{0.711}$	Green and Hall (1975)
$M_R = 18$ CBR$^{0.64}$	Lister and Powell (1987)
$M_R = 21$ CBR$^{0.65}$	Ayres (1997)
$M_R = 17.6$ CBR$^{0.64}$	AASHTO T.307 (2012)
$M_R = 3$ CBR	Edil et al. 2006

Since most of the models over-predicted M_R value, the correlation proposed by AASHTO was adopted in the present study due to its wider acceptance and application. The predicted M_R value for control and lime treated (6% lime addition) case is presented in Fig. 7. From the results, it was observed that the M_R values of the lime treated case are substantially higher than the control case. The M_R values varied from 50 MPa to 198 MPa for CG addition of 10% and 40% respectively. Incoherence with the observations made for CBR values, the highest M_R values were reported for the lime treated case. From the observed CBR and M_R values, it can be inferred that the CG-BC soil mixture (both control and lime treated case) can be utilized as a subgrade/subbase material.

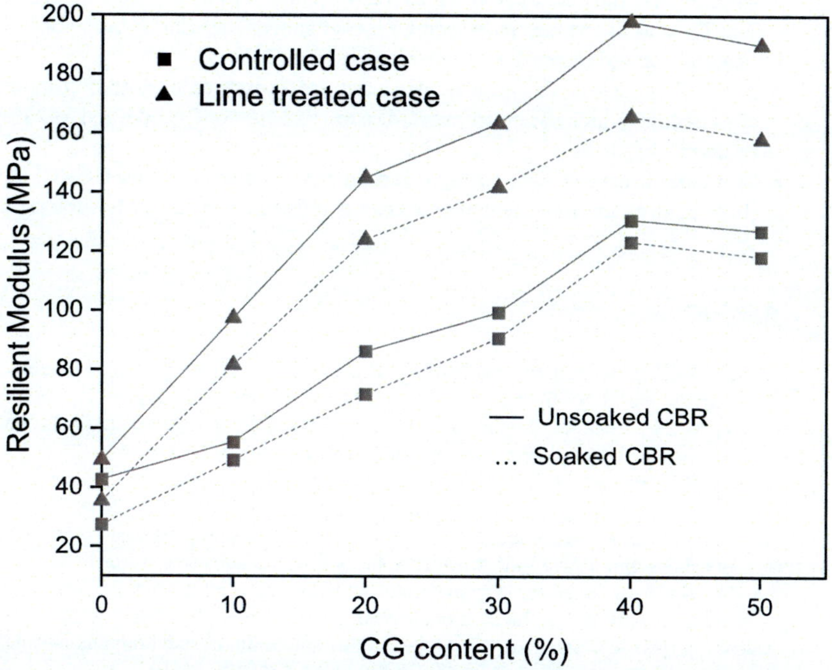

Fig. 7. The variation in M_R of control case and lime stabilized CG-BC soil mixtures.

4 Conclusions

In the current study, the effect of varying quantities (10%–50%) of CG addition on the CBR of Black Cotton (BC) soil is studied. Further, the effect of varying lime content (2%, 4% and 6%) on CBR behavior CG-BC soil mixtures was evaluated, and the following conclusions are drawn:

- The addition of CG to BC soil substantially enhanced the CBR values from 4 to 23 with the maximum values reported for 40% CG addition.

- The lime addition has further increased the CBR values of CG-BC soil mixtures with almost 100% increment for all the combination.
- The curing period has a direct influence on the CBR values of lime stabilized CG-BC soil mixtures with continuous increment observed for curing periods up to 28 days. The maximum CBR values of 4.2 and 3.3 were reported at 28 days of curing period.
- In soaked condition, the CG addition to BC soil has a similar effect on CBR values observed in unsoaked condition. However, the cured samples of lime stabilized CG-BC soil mixtures exhibited relatively lower values under soaking, which is due to lose in surface tension forces.
- The 6% lime dosage with 40% CG addition yielded maximum CBR values for CG-BC soil mixture; thus, it is identified as the optimum combination for stabilization.
- The TM and SM values increased with coal gangue addition in both control and lime treated case and increase is proportional to the CBR values.
- The MR values of the lime treated case are substantially higher compared to the control case, which varied from 50 MPa to 198 MPa for CG addition of 10% and 40% respectively.
- From the observed CBR and M_R values, it can be inferred that the CG-BC soil mixture (both control and lime treated case) can be utilized as a subgrade/subbase material.

References

AASHTO T307: Determining the Resilient Modulus of Soils and Aggregate Materials. American Association of State Highway and Transportation Officials, Washington, DC, USA (2012)

Ashfaq, M., Heeralal, M., Moghal, A.A.B., Murty, V.R.: Carbon Footprint analysis of coal gangue in geotechnical engineering applications. Indian Geotech. J. **50**, 646–654 (2020)

Ashfaq, M., Heeralal, M., Moghal, A.A.B.: Effect of coal gangue particle size on its leaching characteristics. ASCE Geotech. Special Publication **319**, 107–114 (2020)

Ashfaq, M., Heeralal, M., Moghal, A.A.B.: Characterization studies on coal gangue for sustainable geotechnics. Innovative Infrastructure Solutions 5(1), article no 15 (2020b)

ASTM D1883-16: Standard test methods for California bearing ratio (CBR) of laboratory-compacted soils. ASTM International, West Conshohocken, P (2016)

ASTM D4318-17e1: Standard Test Methods for liquid limit, plastic limit and plasticity index of soils. ASTM International, West Conshohocken, PA (2017)

ASTM D698 12e2: Standard Test Methods for Laboratory Compaction Characteristics of Soil Using Standard Effort. ASTM International, West Conshohocken, PA (2012)

ASTM D2487-11: Standard Practice for Classification of Soils for Engineering Purposes (Unified Soil Classification System), ASTM International, West Conshohocken, PA (2017)

ASTM D4972-19: Standard test method for pH of soil. ASTM International, West Conshohocken, PA (2019)

ASTM D854-14: Standard test methods for specific gravity of soil solids by water pycnometer. ASTM International, West Conshohocken, PA (2014)

Ayres, M.: Development of a Rational Probabilistic Approach for Flexible Pavement Analysis. University of Maryland (Publisher), College Park, MD, USA (1997)

Bell, F.G.: An examination of the use of lime and pulverized fly ash to stabilize clay materials. Bull. Assoc. Eng. Geol. **30**(4), 469–479 (1993)

Darikandeh, F.: Expansive soil stabilized by calcium carbide residue–fly ash columns. Proc. Inst. Civil Eng. - Ground Improvement **171**(1), 49–58 (2018)

Edil, T.B., Acosta, H.A., Benson, C.H.: Stabilizing soft fine-grained soils with fly ash. J. Mater. Civ. Eng. **18**(2), 283–294 (2006)

Green, J., Hall, J.: Nondistructive Vibratory Testing of Airport Pavement. Technical Report S-75-14 (1975). Online source: ark:/67531/metadc304084

Heitor, A., Indraratna, B., Kaliboullah, C.I., Rujikiatkamjorn, C., McIntosh, G.W.: Drained and Undrained Shear behavior of compacted coal wash. J. Geotech. Geoenviron. Eng. **142**(5), 04016006, 1–10 (2016)

Heukelom, W., Foster, C.: Dynamic testing of pavements. J. Soil Mech. Found. Division **86**(1), 1–28 (1960)

Indraratna, B., Gasson, I., Chowdhury, R.N.: Utilization of compacted coal tailings as a structural fill. Can. Geotech. J. **31**, 614–623 (1994)

Indraratna, B., Rujikiatkamjorn, C., Chiaro, G.: Characterization of compacted coal wash as structural fill material. ASCE Geotech. Special Publication **225**, 3826–3834 (2012)

Lister, N.W., Powell, D.: Design practices for pavements in the United Kingdom. In: Proceedings of the 6th International Conference on the Structural Design of Asphalt Pavements, Ann Arbor, MI, USA (1987)

Moghal, A.A.B., Chittori, B.C.S., Basha, B.M., Al-Shamrani, M.A.: Target reliability approach to study the effect of fiber reinforcement on UCS behavior of lime treated semiarid soil. J. Mater. Civil Eng. **29**(6), 04017014, 1–15 (2017)

Murthy, V.R., Krishna, P.H.: Stabilization of expansive clay bed using calcium chloride solution. Ground Improvement **10**(1), 39–46 (2006)

Pandian, N.S., Krishna, K.C., Sridharan, A.: California bearing ratio beahvior of soil/fly ash mixtures. J. Test. Eval. **29**(2), 220–226 (2001)

Pandian, N.S., Krishna, K.C.: California Bearing Ratio Beahvior of cement stabilized Fly ash-Soil mixes. J. Test. Eval. **30**(6), 492–496 (2002)

Pandian, N.S., Krishna, K.C.: The pozzolanic effect of fly ash on the california bearing ratio behavior of black cotton soil. J. Test. Eval. **31**(6), 479–485 (2003)

Petri, T.M., Little, D.N.: Review of stabilization of clays and expansive soils in pavements and lightly loaded structures—history, practice, and future. J. Mater. Civ. Eng. **14**(6), 283–294 (2002)

Phanikumar, B.R.: Effect of lime and fly ash on swell, consolidation and shear strength characteristics of expansive clays: a comparative study. Geomech. Geoeng. **4**(2), 175–181 (2009)

Puppala, A.J., Wattanasanticharoen, E., Hoyos, L.R.: Ranking of four chemical and mechanical stabilization methods to treat low-volume road subgrades in texas. Transp. Res. Rec. **1819**(1), 63–71 (2003)

Senol, A., Edil, T.B., Bin-Shafique, M.S., Acosta, H.A., Benson, C.H.: Soft subgrades' stabilization by using various fly ashes. Resour. Conserv. Recycl. **46**, 365–376 (2006)

Tasalloti, S.M.A., Indraratna, B., Rujikiatkamjorn, C., Chiaro, G., Heitor, A.: A laboratory study on the shear behavior of mixtures of coal wash and steel furnace slag as potential structural fill. Geotech. Testing J. **38**(4), 361–372 (2015)

Field Soil Electrical Resistivity Measurements of Some Soil of Iraq

Zuhair Kadhim Jahanger[1](✉), Ali J. Nouri Al-Barazanchi[2], and Azad Abbas Ahmed[3]

[1] Department of Water Resources Engineering, College of Engineering, University of Baghdad, Baghdad, Iraq
zk_jahanger@coeng.uobaghdad.edu.iq
[2] Highway Engineering Department, Erbil Technical College, Erbil Polytechnic University, Erbil, Iraq
ali.nouri@epu.edu.iq
[3] Andrea Engineering Tests Laboratory, Baghdad, Iraq
azad@andrealab.com

Abstract. With the increasing number of power projects comprising water treatment plants and substations in Iraq, the analysis and evaluation of electrical resistivity tests of silty clay soil have become important considerations in engineering construction. The field soil electrical resistivity has not been widely studied yet. However, through field tests for five sites in the middle part of Iraq were performed. Therefore, field electrical soil resistivity tests carried out throughout the site at three different locations in two perpendicular directions. The soil electrical resistivity tests of mostly silty clay and sandy silty clay soils were analyzed. The results showed that the variations in the field soil electrical resistivity can be discussed regarding the soil type and the water table at the time of the test as a time-dependent test. The field soil electrical resistivity decreases rapidly with depth due to the higher water content. Also, the sand content has a significant effect in which the field soil electrical resistivity increases with the increase of the sand. As the plotted resistivities against depth indicate there are not distinct layers of different soil which compared well the site investigation borehole logs. Mostly, a good correlation exists between the in situ soil electrical resistivity and depth suggesting that correlations or fitting curve are reliable to be applied in the future to predict the field soil electrical resistivity in the study area. The results of the field soil electrical resistivity tests performed on the soil verify the soundness of the proposed equation to predict field soil electrical resistivity.

Keywords: Field soil electrical resistivity · In situ · Temperature · Silty clay · Sand

1 Introduction

Geophysical methods are one approach to determine soil properties in addition to laboratory and field (in situ) approaches (Bowles 1997). A geophysical method such as

J. Neves et al. (Eds.): GeoChina 2021, SUCI, pp. 114–122, 2021.
https://doi.org/10.1007/978-3-030-80155-7_8

electrical resistivity is usually measured as electrical parameters in a laboratory and/or field condition (i. e. Atkinson 2007). Field electrical resistivity measurements of subsoil based on electrical resistivity tests including electrode spacing have the advantages of directly measuring the soil electrical resistivity (i.e. Das 2016). Measurement of soil electrical resistivity is very important which is being implemented for the control of corrosion of buried structures such as pipes for the design of cathodic protection systems in different engineering projects such as power plant and water treatment plant. For that, it is important to test as many points as necessary to get a sufficiently representative characterisation of the soil environment. As per the frequency of taking a Wenner four-electrode method resistivity readings could be made regularly at close spacing, except where low-risk areas have been identified by initial site measurements. Spacing between measurements can then be increased (ASTM G57 2015). Locations of the electrical resistivity test will be set out on site, and their coordinates are taken, during the survey services carried for the site as well as during the field test. The experience of the engineer concerned will be made a significant effect on the results of the field soil resistivity measurements. Soil electrical resistivity has been investigated in a large number of studies (i.e. Gance et al. 2016), which can be correlated. However, from user experience, interpretation of the geophysical method results will help to develop a relationship between the in situ soil resistivity and depth that are reliable to be applied in the future to predict the field soil electrical resistivity of same soil properties.

In the literature, many correlations proposed between soil resistivity and soil properties (Ozcep et al. 2009). Most of the models depend on a series of laboratory experiments (Bery and Ismail 2018; Seladji et al. 2010). Laboratory electrical resistivity tests may be employed in laboratory investigations and these should be defined in reporting the results. The assessment of soil water content, density and soil type regarding the variations of soil electrical resistivity with depth depending on the geotechnical engineering experience is very important and required to add an acknowledgement to the end-users of the test results. Detailed information about the existing field soil electrical resistivity shall be evaluated in the vicinity of new results in this regard that have to be established and understood physically.

The study aimed to assess the relationships between field soil electrical resistivity and soil type (grain size) with a water table (W.T.) as well as the effect of the overall soil electrical resistivity behaviour of mainly of silty clay and/or silty sand layers resting over (clayey) silty sand using the Wenner four-electrode method. Moreover, an initial analysis was performed to develop a practically applicable correlation for the determination of field soil electrical resistivity for the soil type used here for further study in the future. The focus of this investigation was limited to study area in Babylon Governorate in Iraq of sedimentary soil.

2 Subsurface Ground Conditions of the Study Area

The study areas were located in the middle part of Iraq in alluvial plain as shown in Fig. 1. The climate is hot and dry influenced by the subtropical aridity of the Arabian desert areas and the subtropical humidity of the Gulf. The purpose of the investigations was to explore the subsoil conditions of the proposed sites for the construction of the structures and facilities comprising buried pipelines for water treatment project and power substations. For geotechnical purposes, during fieldwork, one borehole 12 m depth was drilled for this study at each site all from the existing natural ground level (ASTM D1452). The fieldwork was conducted in fall-winter in 2019 including the execution of standard penetration test (ASTM 1586) and collected undisturbed samples using thin wall Shelby tube according to (ASTM D-1587). The related physical, mechanical and chemical characteristics of the subsoil were measured (Jahanger et al. 2018; Jahanger 2021) on the obtained soil samples (Fig. 2). The results of the subsoil of the site that are derived from the field and laboratory testing performed on site's samples consist mainly of silty clay and/or silty sand layers resting over clayey sand and/or silty sand. However, the W.T. encountered at depths of about 3.0 m below the existing ground level (NGL) for site 1–4, and 0.7 m for site 5 conducted following ASTM D4750, which is considered in shallow range of depth (i.e. Powrie 2014). The W.T. observed to fluctuate with the seasons in the study area in which can be rising during Spring. Therefore, the soil immediately above the W.T. is significantly affected as far as compressibility and strength are concerned. The former increases, and the latter decreases with an increase in the water content of the soil. Soluble salts association with organic matter were found in the soil in several forms in a significant amount up to ≈6%.

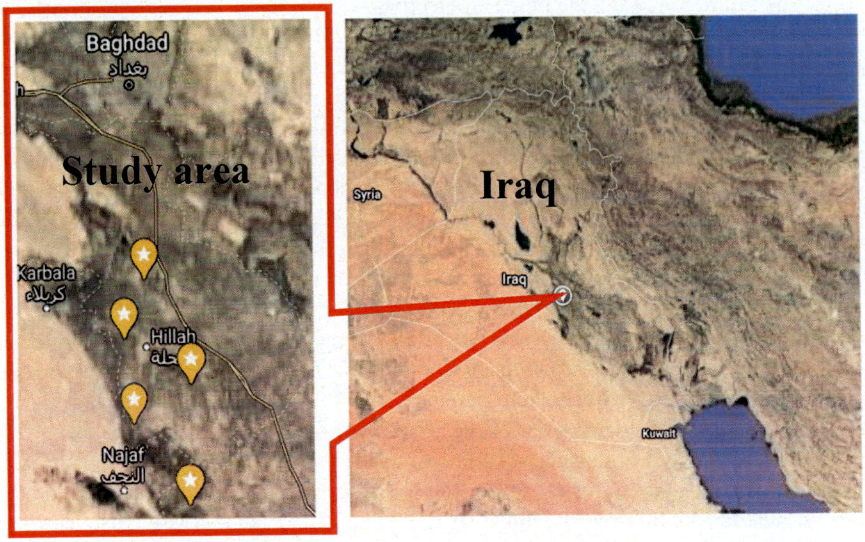

Fig. 1. Location map of the study area

For comparison with the soil resistivity for all the sites, the results of the field ground temperature measurements with depth is studied in as shown in Fig. 3 using a portable TLS-100 m (Thermtest Inc). There seems to be a pattern and a difference in ground temperature due to the effects of different types of soil. The difference in the value of measured ground temperature between the soils increases and becomes relatively lower with depth as approaching the W.T. But ground temperature profile for site 5 increases

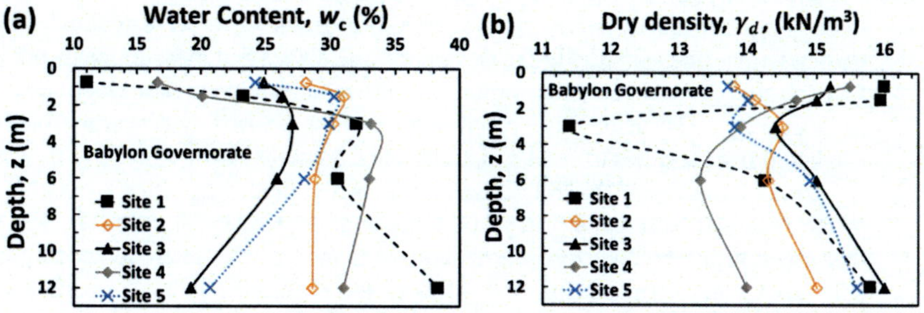

Fig. 2. (a) Water content and (b) Dry density with depth

with depth at an increasing rate as a homogenous continuum stratum where W.T. is 0.7 m which heating time should be reduced for soils with high moisture contents. It appears that the ground temperature profile remains almost constant between 25–28 degree centigrade below the W.T. if extrapolated from the current data. Moreover, though the data is not included here, it is worth mentioning that at a shallower depth, the ground temperature at a certain depth decreases rapidly with time and at a deeper depth, the temperature slightly decreases with time.

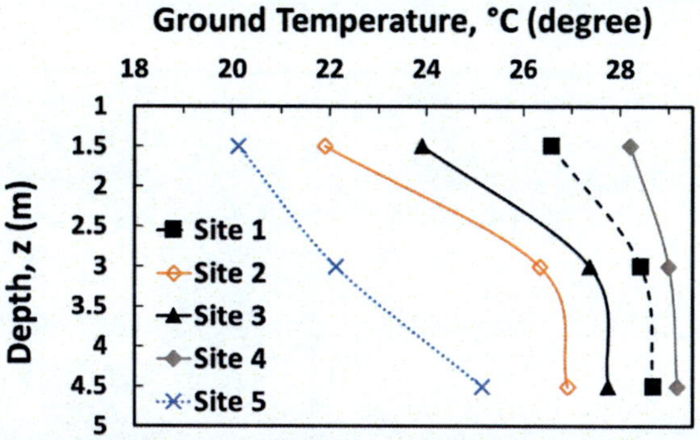

Fig. 3. Subsoil temperature versus depth

3 Electrical Resistivity Test

The electrical resistivity of soil is defined as the electrical resistance of a unit cubic of the soil (ASTM G57). The electrical resistivity of soil is required for designing a safety grounding system. The electrical set-up used in this study is depicted in Fig. 4. Electrical resistivity is measured by the SR-2 Model, by Tinker and Rasor. Wenner's four-electrode method as specified in IEEE standard 81 and ASTM G57 was used. The four-point resistivity tests were performed using four metal probes (electrodes) be set in a straight line in the surface with equal spacing between the probes. At each site, three field electrical resistivity tests at various locations tests were conducted along two perpendicular lines parallel to the site's coordinate axes and measurements are averaged for each site. On each line, measurements were taken at incremental electrode spacing of (*a*) in m (Fig. 4). By using this technique, an apparent resistivity 1D vertical depth-profile is created from a series of field resistivity measurements.

By applying Ohms law, the resistance (R) was calculated, which is equal to the ratio of the potential difference (ΔV) to the applied current (I) (R = ΔV/I). The depth of investigation is obtained by progressively expanding the electrode spacing (*a*), from 0.75 m, 1.5 m, 3.0 m, 6.0 m and 12 m to the required depth of investigation. Most pipelines are installed at shallow depths of from 1.5 to 4.5 m, electrode spacing of 1.5, 3.0 and 6.0 m are commonly used according to ASTM G-57. However, the spacing of the electrodes (*a*) should equal the maximum depth of interest. Therefore, the depth of investigation for each electrode configuration is equal to approximately a third to half of the electrode spacing (*a*/3 to *a*/2). Also, the measurement depth is usually assumed to be equal to the distances between the electrodes (Samouëlian et al. 2005). To obtain apparent resistivity values, a geometric factor unique to the electrode spacing and array must be applied to the resistance data. Resistivity measurements are usually controlled by the main soil type in terms of its chemical composition, grain size and shape, and more importantly by the moisture content of the same soil at the same depth. The resistivity measurements are usually very useful in many disciplines such as environmental and civil engineers.

Fig. 4. Soil resistance meter, model SR-2, by Tinker and Rasor

With the resistance meter, the earth's resistance or ΔV/I is automatically calculated and displayed in Kilo-ohm (KΩ), ohm (Ω) or in milliohm (mΩ). By applying the proper geometrical factors to the measured resistance, the apparent resistivity can be calculated for each measurement, and usually expressed in Ohm meter (Ωm). Both lateral and vertical variations in apparent resistivity can be detected by either adopting vertical electrical sounding (VES) or electrical resistivity tomography (ERT) mode of operation. The calculated resistivity for uniform soil layer is constant and independent of both the electrode spacing and surface location (Samouëlian et al. 2005). In this study, the apparent resistivity ρ (Ωm) is calculated for each electrode spacing using the equation:

$$\rho = 2\pi a \mathrm{R} \tag{1}$$

To assess the soil corrosivity, reference is made to the British Standard BS7361: Part 1: 1991, in which measured apparent resistivity values are related to soil corrosivity as in Table 1.

Table 1. Apparent resistivity values to soil corrosivity

Apparent resistivity, Ωm	Corrosivity index
up to 10	Severely corrosive
10 to 50	Corrosive
50 to 100	Moderately corrosive
100 and above	Slightly corrosive

4 Field Soil Electrical Resistivity Results

In this study, the locations of the field electrical resistivity test were set out on site. The measured apparent resistivity values are plotted as a function of the electrode spacing (a) on a normal scale to produce a resistivity sounding curve as illustrated in Fig. 5. Quantitative interpretation of the resistivity sounding curves were carried out using a regression equation. For each site, a correlation curve can be deduced, showing the apparent resistivity and thickness of the various subsurface layers, together with the error of fit for this interpreted regression equation. It can be seen that the measured soil resistivity values decrease as the grain size decreases as the soil changes from sand (coarse-grained) to clay (fine-grained), as shown in Fig. 5. In general, this is obvious in Fig. 5 (embedded arrow) where it shows that the sandy soil (coarse-grained) has a higher electrical resistivity than clayey soil (fine-grained).

Soil size (grain size) in both the saturated (below W.T.) and unsaturated (above W.T.) zones, is the most affecting factor on the field soil electrical resistivity as illustrated in Fig. 5. The average measured field electrical resistivity on the silty sand soil for different densities as a function of electrode spacing (a) decreased when the content of fine-grained soil (clay) increased. This decrease was the smallest at the shallowest water table (Site 5) and with the smallest water contents. Also, the W.T. has a significant effect on the average measured electrical resistivity which could cause a decrease of 60% of the maximum measured soil electrical resistivity that measured at the water table. As can be seen in Fig. 5 that below the W.T. the measured soil electrical resistivity reaches a plateau which is not related to the moisture content (Fig. 2a).

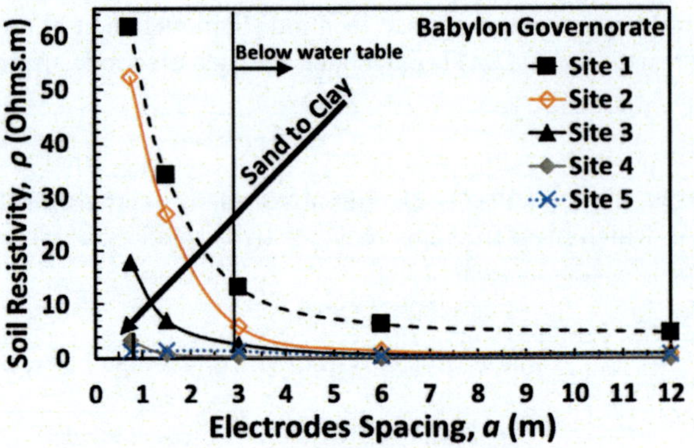

Fig. 5. Soil resistivity measurement with distance

5 Discussion

As observed in Fig. 6, the field soil electrical resistivity ($\boldsymbol{\rho}$) varies with the depth of the soil layer. Therefore, a relation between ρ and a from the field measures (Fig. 6) has been presented in Eqs. (2–6). As expected, the plots of the average soil electrical resistivity for different a, show that $\boldsymbol{\rho}$ is variable and depending on soil grain size and the water table (Fig. 5). For comparison purposes and according to Seladji et al. 2010, it is worth mentioning that soils with a greater percentage of fines or clay, or a smaller coarse fraction have lower soil electrical resistivity. Therefore, from the field test data used in Fig. 6, a power, logarithmic and third-order polynomial equations were obtained, as it was the best fit using the regression analysis as follows:

$$\text{For site 1}: \quad \rho = 45.069a^{-0.986}, R^2 = 0.97 \tag{2}$$

$$\text{For site 2}: \quad \rho = 37.235a^{-1.644}, R^2 = 0.97 \tag{3}$$

$$\text{For site 3}: \quad \rho = 10.802a^{-1.25}, R^2 = 0.91 \tag{4}$$

$$\text{For site 4}: \quad \rho = -1.055\ln(a) + 2.0265, R^2 = 0.71 \tag{5}$$

$$\text{For site 5}: \quad \rho = 0.0093a^3 - 0.15a^2 + 0.47a + 1.015, R^2 = 1 \tag{6}$$

For comparison purposes, it is worth mentioning that the soil electrical resistivity is not significantly related to the water content that much for the saturation soils (below W.T.) as shown in Fig. 5 and Fig. 2a. The measured field water content either increased or decreased with depth while the soil electrical resistivity decreased. In addition to that, the soil electrical resistivity is not related to the dry density in such a way that can be described properly (Fig. 2b). This result was contradicted by Seladji et al. 2010. They reported that the field soil electrical resistivity of compacted clay is sensitive to compaction conditions, with lower electrical resistivity obtained for compaction at

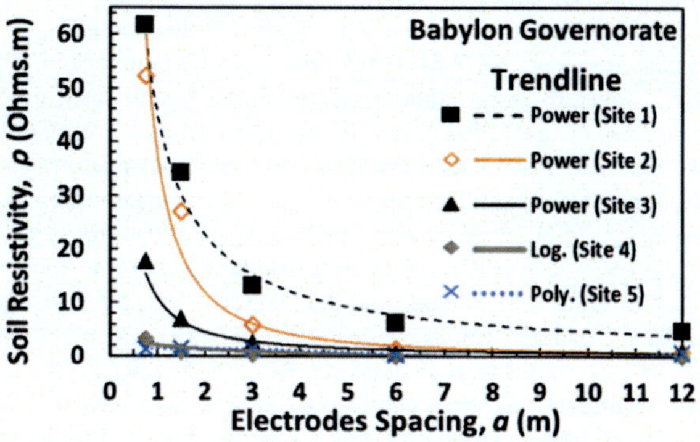

Fig. 6. Soil resistivity measurement with distance

higher water content or greater compaction effort. However, it can be seen that the soil electrical resistivity related to the water table at the time of the test (Fig. 5). It seems that the compilation effect of water and voids.

6 Conclusions

This paper has presented the field electrical resistivity results for some Iraqi soils aimed at studying the actual response of silty clay and silty sand in terms of water content, soil size and dry density. In this study, water content either increased or decreased while the soil electrical resistivity rapidly decreased with depth. Also, at a shallower depth, the subsoil temperature at certain depth decreases rapidly with time and at a deeper depth, the temperature slightly decreases with time; however, subsoil temperature increases with an increase in the depth. It appears that below the W.T., the ground temperature profile remains almost constant between 25–28° centigrade. The soil electrical resistivity drops significantly at the water table. These observations can suggest that the W.T. has a significant effect on the soil electrical resistivity. Moreover, it is worth mentioning that soils with a greater percentage of fines or clay, or a smaller coarse fraction have lower soil electrical resistivity. It has been confirmed that the field soil electrical resistivity is significantly related to the grain size and water table. A good correlation (fitting curve) exists between the field soil electrical resistivity and depth suggesting that methods are reliable to be applied in the future to predict the field soil electrical resistivity in the study area. Finally, based on this study, the obtained field soil electrical resistivity measurements could reflect true resistivity. These can be used without limitation and provide useful measurements from the macroscopic to field scale.

References

ASTM D1452. Standard Practice for Soil Exploration and Sampling by Auger Borings. ASTM International, West Conshohocken, PA (2016)

ASTM 1586. Standard Test Method for Standard Penetration Test (SPT) and Split-Barrel Sampling of Soils. ASTM International, West Conshohocken, PA (2018)

ASTM D1587. Standard Practice for Thin-Walled Tube Sampling of Fine-Grained Soils for Geotechnical Purposes. ASTM International, West Conshohocken, PA (2015)

ASTM D4750. Standard Test Method for Determining Subsurface Liquid Levels in a Borehole or Monitoring Well (Observation Well). ASTM International, West Conshohocken, PA (1989)

ASTM G57. Standard Test Method for Field Measurement of Soil Resistivity Using the Wenner Four-Electrode Method. ASTM International, West Conshohocken, PA (2012)

ASTM: Soil and rock, building, stores, geotextiles. American Society for Testing and Materials, ASTM International, West Conshohocken, PA, **04.08** (2015)

Atkinson, J.: The Mechanics of Soils and Foundations, 2nd edn. CRC Press, London (2007)

Bery, A.A., Ismail, N.E.H.: Empirical correlation between electrical resistivity and engineering properties of soils. Soil Mech. Found Eng. **54**(6), 425–429 (2018)

Bowles, J.E.: Foundation Analysis and Design, 5th edn. McGraw-Hill, Singapore (1997)

BS 7361-1: 1991: Cathodic Protection, Part 1: Code of Practice for Land and Marine Applications (1991)

Das, B.M.: Principles of Foundation Engineering, 8th edn. Cengage Learning, India (2016)

IEEE Standard 81-1983: IEEE Guide for Measuring Earth Resistivity, Ground Impedance, and Earth Surface Potentials of a Ground System. New York, USA (1983)

Gance, J., Malet, J.-P., Supper, R., Sailhac, P., Ottowitz, D., Jochum, B.: Permanent electrical resistivity measurements for monitoring water circulation in clayey landslides. J App Geophy (2016). https://doi.org/10.1016/j.jappgeo.2016.01.011

Jahanger, Z.K.: Evaluation of the thermal conductivity of middle part of Iraqi soil. Mater. Today: Proc. **42**, 2431–2435 (2021)

Jahanger, Z.K., Sujatha, J., Antony, S.J.: Local and global granular mechanical characteristics of grain–structure interactions. Indian Geotechn. J. **48**(4), 753–767 (2018). https://doi.org/10.1007/s40098-018-0295-5

Ozcep, F., Tezel, O., Asci, M.: Correlation between electrical resistivity and soil-water content: Istanbul and Golcuk. Int. J. Phy. Sci. **4**(6), 362–365 (2009)

Powrie, W.: Soil mechanics: concepts and applications, 3rd edn. CRC Press, London, UK (2014)

Samouëlian, A., Cousin, I., Tabbagh, A., Bruand, A., Richard, G.: Electrical resistivity survey in soil science: a review. Soil Tillage Res. **83**(2), 173–193 (2005)

Seladji, S., Cosenza, P., Tabbagh, A., Ranger, J., Richard, G.: The effect of compaction on soil electrical resistivity: a laboratory investigation. Eur. J. Soil Sci. **61**(6), 1043–1055 (2010)

Author Index

J. Neves et al. (Eds.): GeoChina 2021, SUCI, p. 123, 2021.
https://doi.org/10.1007/978-3-030-80155-7